JOSEPH LAU

STATISTICAL SIGNIFICANCE

STATISTICAL SIGNIFICANCE

Rationale, Validity and Utility

Siu L. Chow

SAGE Publications
London • Thousand Oaks • New Delhi

© Siu L. Chow 1996

First published 1996

SAGE Publications Ltd
6 Bonhill Street
London EC2A 4PU

SAGE Publications Inc
2455 Teller Road
Thousand Oaks, California 91320

SAGE Publications India Pvt Ltd
32, M-Block Market
Greater Kailash - I
New Delhi 110 048

British Library Cataloguing in Publication data

A catalogue record for this book is
available from the British Library

ISBN 0 7619 5204 7
ISBN 0 7619 5205 5 (pbk)

Library of Congress catalog card number 96-070970

Typeset by Paston Press Ltd, Loddon, Norfolk
Printed in Great Britain by Biddles Ltd, Guildford, Surrey

Contents

Preface

To conduct empirical research is to engage in an exercise which requires conceptual, theoretical and statistical skills. The ultimate goal is to collect and analyse empirical data in a systematic way so as to give valid answers to well-defined research questions. Consequently, the validity (or correctness) of empirical studies is assessed in terms of the aforementioned three kinds of expertise. In practice, however, the assessment of research results is complicated by the fact that researchers conduct empirical studies for various reasons – for example, to corroborate explanatory theories, to identify or categorize phenomena, to establish functional relationships among variables, to ascertain the generality of established functional relationships among variables, to assess the effectiveness of intervention programmes or to evaluate the practical impact of research findings.

Researchers may weigh the relative importance of the conceptual, theoretical and statistical components of empirical research differently in accordance with their research objectives. This may be illustrated as follows. Suppose that Study A is conducted to test an explanation of hyperactivity. Specifically, the aim of Study A is to examine *why* a stimulant drug alleviates hyperactivity symptoms. The purpose of Study B, on the other hand, is to ascertain *whether or not* a stimulant drug is efficacious in reducing hyperactivity symptoms. It may be readily seen that Studies A and B are to be assessed with different criteria.

To begin with, it is more important to exclude alternative interpretations of data at the conceptual level in the case of the *why* objective than of the *whether or not* objective. Hence, conceptual or theoretical considerations are more important in the case of Study A than Study B as a result of the need to exclude alternative conceptual interpretations of data in Study A. Pragmatic considerations, on the other hand, are more important for Study B. The utilitarian objective of the research question may be achieved at the expense of conceptual rigour. It may not be important to determine why the stimulant drug works in the case of Study B. In other words, conceptual or technical considerations important to Study A may be deemed unnecessary or unrealistic to the researcher who conducts Study B.

Of concern is that, although the null-hypothesis significance-test procedure (NHSTP) is an integral component of data analysis in empirical research, many researchers have reservations about its validity or utility. The task of assessing the role of the NHSTP in empirical research is complicated by the facts that (a) researchers use different research methods to achieve different goals, (b) the interrelations among the conceptual, theoretical and statistical constraints on researchers differ when the researchers have different research objectives, and (c) the relationship between the substantive and statistical hypotheses is often oversimplified, if not ignored altogether, in meta-theoretical discussions.

The thesis of this book is that NHSTP has been criticized for the wrong reasons. It will be shown that the critics of the procedure (called 'critics' in subsequent discussion) find NHSTP wanting because, for various reasons, they have attributed to it functions for which statistics cannot, nor should be expected to, serve. Specifically, there are more doubts about NHSTP (a) when researchers are less concerned with theoretical considerations, or (b) when it is more difficult to exclude alternative interpretations of research results. It will be argued that NHSTP, as a statistical decision procedure, plays only a very limited, although a valid and indispensable, role in empirical research. The more positive purpose of defending NHSTP is to fill a pedagogical gap.

Quite a few meta-theoretical issues and conceptual distinctions are implicated in a thorough examination of the rationale, validity and utility of NHSTP. However, these meta-theoretical issues and conceptual distinctions are discussed neither in introductory statistics textbooks nor in research methods textbooks. One reason for this neglect may be that the pioneers of statistics known to researchers in the social sciences (viz., K. Pearson, Fisher, Neyman and E.S. Pearson) did not distinguish between statistical and non-statistical issues when they engaged in methodological discussions. A second reason may be that contemporary authors of research methods textbooks find it inappropriate to deal with these issues and distinctions at the introductory level.

However, these meta-theoretical issues (e.g., the nature and methodological utility of inductive methods) and conceptual distinctions (e.g. *theory-corroboration* versus *statistical hypothesis testing*) become important in view of the recent attempts by critics to exhort empirical researchers (a) to report confidence intervals rather than only statistical significance, (b) to base their research conclusions on the estimate of confidence interval or effect size rather than on statistical significance, and (c) to select the sample size with reference to statistical power. These issues and distinctions are also important when the validity of meta-analysis is examined or when the Bayesian criticisms of NHSTP are considered.

This book begins with a summary of various criticisms of NHSTP in Chapter 1. The hybridism, rationale, procedure and objective of NHSTP are introduced in Chapter 2. It will be shown that the null hypothesis is never used as a categorical proposition. The implication of this state of affairs plays an important role in this defence of NHSTP. It will also be shown that the probabilistic basis of NHSTP is a sampling distribution of the test statistic. Moreover, a case may be made that probability statements in empirical research are statements about data, not hypotheses.

Chapter 3 describes the interrelations among (a) the to-be-studied phenomenon, (b) the to-be-corroborated theory or substantive hypothesis, (c) the research hypothesis derived from the substantive hypothesis and (d) the experimental hypothesis specified in a particular context. These interrelations are described with reference to the rationale of the theory-corroboration experiment. The statistical alternative hypothesis is the implication of the experimental hypothesis at the statistical level. NHSTP is defended in the context of experimentation because two prominent critics of NHSTP have made the point that their criticisms of NHSTP are directed to non-experimental studies.

The logical foundation (both deductive and inductive) of the theory-corroboration experiment is introduced in Chapter 4. This logical foundation serves as the frame of reference for assessing the meta-theoretical issues and conceptual distinctions implicated in some criticisms of NHSTP. It will be shown that there is more to induction than enumeration plus generalization.

Difficulties with supplementing, if not replacing, NHSTP with estimates of the effect size are discussed in Chapter 5. Also examined are the difficulties of treating meta-analysis as a theory-corroboration procedure. Chapter 6 is devoted to an examination of the power of a statistical test. Some researchers claim that Bayesian statistics is to be preferred to NHSTP. The meta-theoretical assumptions underlying Bayesian statistics are examined in Chapter 7. A summary of the main arguments in defence of NHSTP is presented in Chapter 8.

To K. Pearson, R. Fisher, J. Neyman and E.S. Pearson, NHSTP was what the empirical research method was all about. At the same time, many of their meta-theoretical considerations about NHSTP (and, by extension, about the research process itself) were not the fruit of a systematic reflection about the research process. Instead, many of the meta-theoretical statements came about as the result of the less than cordial exchanges among K. Pearson, Fisher, and the Neyman–E.S. Pearson duo. The situation became more complicated when the disputants sometimes changed positions, or incorporated their opponent's ideas as their own with no explicit acknowledgment to that effect (see Gigerenzer, 1993; Inman, 1994; Oakes, 1986). This may be responsible for the hybridism of NHSTP identified by Gigerenzer (1993).

This state of affairs may also be responsible for the fact that contemporary criticisms of NHSTP are carried out in the absence of a coherently developed frame of reference. Be that as it may, there is a positive way to look at such criticisms. The critiques are attempts to challenge NHSTP users to rationalize the research procedure in general, and the role of NHSTP in such a procedure in particular. The present defence of NHSTP is an attempt to meet the challenge. This account of NHSTP is not expected to be the final one. None the less, it will fulfil its purpose if it serves as the basis for further exploration of the issues raised in the course of the present argument. It is hoped that a coherent view of NHSTP pertinent to empirical research will emerge from the ensuing discussion.

I thank Christopher Essex, Robert Frick, George Maslany, Richard MacLennan, Jeffrey Pfeifer, Kenneth Probert and William Smythe for their help or comments on various sections of the book.

1

A Litany of Criticisms of NHSTP

How NHSTP is applied in a one-factor, two-level experiment is used as a backdrop for documenting some well-known criticisms of NHSTP. Tests of significance are faulted because both the null and alternative hypotheses are considered problematic. A statistically significant result is ambiguous because of the arbitrary features found in NHSTP. Statistical significance is not informative as to the effect size, the probability that the theory is true or the practical importance of the result. Researchers may readily be misled by statistical significance. Psychologists' reliance on NHSTP leads to a methodological paradox. The ultimate difficulty in using NHSTP is the fact that the null hypothesis is never true.

1.1. Introduction

Many criticisms of the null-hypothesis significance-test procedure (NHSTP) have been made since the 1960s (Morrison & Henkel, 1970). Researchers have long been advised to report confidence-interval estimates or estimates of the effect size in addition to statistical significance. A more recent sentiment is that psychologists should abandon the use of NHSTP altogether (Schmidt, 1994; 1996). However, the emphasis in contemporary statistics textbooks is still on statistical significance. Although statistical power is now defined and given a cursory explanation in introductory textbooks, there is yet no attempt to show neophytes how to calculate statistical power. The situation is deemed so unsatisfactory that a primer on statistical power is now provided (J. Cohen, 1992b) in addition to the more advanced text (J. Cohen, 1987). Despite these efforts, statistical power is still not as universal a concern among empirical researchers as power analysts deem warranted. At the same time, many researchers persist in relying on statistical significance, despite the argument that NHSTP is not a good tool to use to justify scientific knowledge (Rosnow & Rosenthal, 1989; Rozeboom, 1960).

To facilitate the documentation of criticisms of NHSTP, an independent-sample one-factor, two-level experiment is first described. It provides the context for describing the various difficulties engendered as a result of using NHSTP. These difficulties will be grouped into several broad categories: (a) the debatable nature of the null (H_0) and alternative (H_1) hypotheses; (b) the difficulties with the concept statistical significance; (c) questions about the level of significance α, and the associated probability of the test statistic, p; (d) what NHSTP cannot do; (e) the arbitrary and anomalous features found in NHSTP; and (f) some questionable consequences of using NHSTP. Also important is the paradoxical consequence of using NHSTP identified by Meehl (1967).

1.2. A Student-*t* Test Application

Suppose that Psychologist P has good reasons to expect that a new method of teaching statistics (Method E) is superior to the orthodox method (Method C). To assess the putative superiority of Method E to Method C, 20 introductory psychology students are split randomly into two groups of equal size. Groups I and II are taught statistics by Methods E and C, respectively. Both groups are given a test at the end of a 10-week period. For each group of students, the mean and standard deviation of their test performance are calculated. The difference between the two group means (i.e., $\overline{X}_E - \overline{X}_C$) and the *standard error of the difference between two means* for the two independent groups are also obtained. Consequently, the independent-sample t statistic can be calculated.

Assume further that the value of the associated probability of the test statistic, p, is also available. At the risk of stating the obvious, the steps implicated in this one-tailed application of NHSTP are shown in Table 1.1. The calculated t (viz., 2.05) exceeds the critical value of $t = 1.734$ (with df = 18). Hence, the statistical decision is to reject H_0 in favour of H_1. As the

Table 1.1   *The sequence of events implicated in testing a statistical hypothesis*

1. Adoption of $\alpha = .05$			
2. Setting of hypotheses:	$H_0: u_E - u_C \leq 0$		
	$H_1: u_E - u_C > 0$		
3. The numerator of the test statistic		$(\overline{X}_E - \overline{X}_C) = 11$	
4. The standard error of the difference	5.37	or	7.05
5. Calculated $t = 2.05$; ($p = .037$) (left panel of Figure 1.1)		or	Calculated $t = 1.56$; ($p > .05$) (right panel of Figure 1.1)
6. df = 18			
7. Critical t (df = 18) = 1.734			Critical t (df = 18) = 1.734
8. Decision:	Calculated $t \geq$ Critical t; reject H_0	or	Calculated $t <$ Critical t; do not reject H_0

observed significant difference (at the .05 level) is in the predicted direction (viz., $u_E > u_C$), the experimental conclusion is drawn that Method E is superior to Method C.

That the decision is to reject H_0 does not mean that H_0 is necessarily false. In the event that H_0 is true, rejecting H_0 is a Type I error. The probability of committing a Type I error (viz., α) is represented by the area shaded with horizontal lines in both panels of Figure 1.1. The associated probability of the calculated t (i.e., p) is graphically represented by the area shaded with vertical lines in the figure. As may be seen from the left panel of the figure, the associated probability of $t = 2.05$ is smaller than the α level.

As another example, suppose that the calculated t is 1.56, which is smaller than the critical t (viz., 1.734). The associated probability of $t = 1.56$ is larger than the α level (see the right panel of Figure 1.1). The decision would be to accept H_0 (or not to reject H_0). However, that does not mean that H_0 is necessarily true. A Type II error is committed if a false H_0 is not rejected.[1]

With the commonly known elements of NHSTP in place, it is possible to recapitulate the litany of criticisms of NHSTP.

1.3. The Null and Alternative Hypotheses

A directional null hypothesis is used in this example. This directional null hypothesis is, in fact, made up of two hypotheses: (a) $H_{0'}$: $u_E - u_C < 0$ (called 'negative-null' for ease of exposition), and (b) H_0: $u_E - u_C = 0$ (called 'point-null' by Meehl, 1967). The point-null hypothesis is more commonly known as 'a hypothesis of no difference', a name clearly not true of the negative-null hypothesis. However, it can readily be shown that whatever is said about the rejection of the point-null hypothesis can be said about the rejection of the negative-null hypothesis, a point to be developed in Section 2.2.5 below (see also Meehl, 1967).

What is asserted in the point-null hypothesis is that there is no difference between two population means. When stated in such bold terms, the null hypothesis is dismissed at once by critics because it is inconceivable to them that two populations in the real world can be identical. In fact, it is self-evident to many critics of NHSTP that the null hypothesis is never true (Bakan, 1966; Binder, 1963; J. Cohen, 1994; Grant 1962; Greenwald, 1993; Lykken, 1968; Meehl, 1967; Nunnally, 1960; Oakes, 1986; Pollard, 1993; Schmidt, 1992; Serlin & Lapsley, 1985, 1993). Hence, NHSTP is found wanting because to reject something that is never true to begin with is not much of an accomplishment.

There is also another complication. Bakan (1966) suggested that a null hypothesis is the to-be-nullified hypothesis. This characterization suggests that the researcher may use the 'null hypothesis' designation in a way that is consistent with the experimenter's vested interests. This state of affairs undermines the integrity of NHSTP as a research tool.

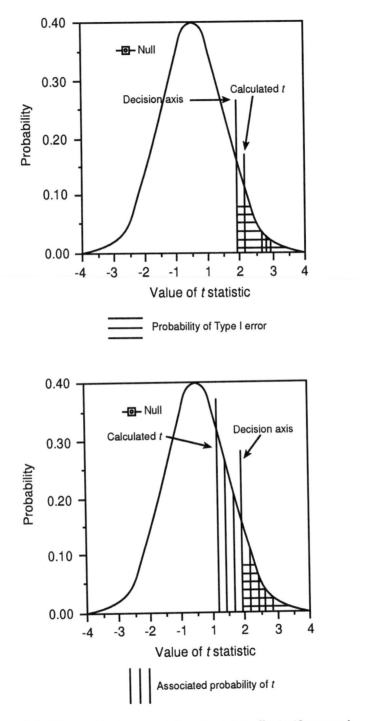

Figure 1.1   *The graphical representation of a statistically significant result (upper panel) and a statistically non-significant result (lower panel)*

It is a conventional practice to identify the hypothesis of interest with the statistical alternative hypothesis. As may be seen from the example in Section 1.2, the statistical alternative hypothesis is one about an ordinal relationship (e.g., H_1: $u_E - u_C > 0$). It is not specific enough at the quantitative level (e.g., H_1: $u_E - u_C = 5$). That is, H_1 says nothing about the exact magnitude of the predicted difference. To some critics (e.g., Meehl, 1967), this is an indication that psychological theories are not sufficiently well developed at the quantitative level to generate point-predictions (e.g., H_1: $u_E - u_C = 5$).

Arguing that a quantitative H_1 is more desirable, critics also note that there is no reason why there is only one H_1; in fact, multiple H_1's should be tested (Neyman & Pearson, 1928; Rozeboom, 1960). For example, instead of predicting H_1: $u_E - u_C \neq 0$, a researcher may predict (a) H_1: $u_E - u_C = 5$, (b) H_1: $u_E - u_C = 10$, (c) H_1: $u_E - u_C = 15$, and so on. Rejecting H_0 is not informative as to which of the multiple H_1's is true.

1.4. Statistical Significance and the Level of Significance

Statistical significance refers to the rejection of the null hypothesis at a predetermined probability of Type I error (α). The first difficulty is that the choice of the α level is arbitrary (Glass, McGaw & Smith, 1981). A consequence of this arbitrariness is the following anomaly. The calculated t in our example (2.05) is significant at the .05 level. However, it would not be significant had the α level been .01 because the critical value of t (with df = 18) is 2.552 at the .01 level. That is, statistical significance is not a stable property or characteristic of the data. Rather, it depends on an arbitrary criterion, namely, the α level. It is no wonder that NHSTP is found wanting as a means to pass judgement on scientific theories (Rozeboom, 1960).

Even if this anomaly could be ignored, critics still would not find it useful to know merely that the result is statistically significant because statistical significance is not informative as to the probability of H_1 being true (J. Cohen, 1987). Nor is statistical significance informative of the true value of the parameter (e.g., u). Furthermore, knowing that the result is significant at the α level is not informative as to the associated probability of the test statistic, p, (Pollard, 1993).[2] It is for these reasons that many researchers suggest replacing tests of significance with confidence-interval estimates (Bakan, 1966; J. Cohen, 1990; Grant, 1962; Lykken, 1968; Meehl, 1967; Nunnally, 1960; Rozeboom, 1960). Furthermore, statistical significance does not provide any information about the substantive or practical importance of the research (Rosenthal, 1983; Rosnow & Rosenthal, 1989; Wilson, Miller & Lower, 1967).

Suppose that statistical significance is not reached. What this really means is not clear to some critics. First, the non-significance may be due to the fact that the test used is not of sufficient power (J. Cohen, 1987). Second, although it is clear in NHSTP that the null hypothesis is to be rejected if the results are significant, the acceptance of the null hypothesis is not inevitable when the results are insignificant (Rozeboom, 1960) because the associated probability, p, is not the probability of H_0 being true.

A further putative limitation of carrying out NHSTP is recognized with the advent of meta-analysis. Merely reporting that the result is significant at the α level is said to be not as helpful for meta-analysts (for example, Glass et al., 1981; Rosenthal, 1984) as reporting the p value or, better still, the actual effect size.

1.5.   The Relative Importance of Type I and Type II Errors

Type I and Type II errors are inversely related. To critics, researchers seem to ignore Type II errors while paying undue attention to Type I errors because of the researchers' concern for statistical significance. This betrays the assumption entertained by NHSTP users that the Type I error is more important than the Type II error. (Schmidt, 1992, for example, believes that NHSTP users entertain this incorrect assumption.) Furthermore, research results ultimately have a bearing on pragmatic actions for which the consideration of Type II error is important. Yet, there is no provision in NHSTP for the determination of Type II error (Bakan, 1966). As far as critics are concerned, researchers are thereby deprived of the important information about the utility of the research results (Rozeboom, 1960).

1.6.   Misinterpretations and a Limitation of the Associated Probability (p)

The associated probability, p, of the test statistic is open to misinterpretation. First, the associate probability is not an exact probability. That is, p is not the exact probability of obtaining a particular value for the test statistic. (For example, .037 in Table 1.1 is not the exact probability of obtaining $t_{(df=18)} = 2.05$.) This is the case because p is 'the probability of a particular value *plus* the probabilities of all more extreme possible values' (Siegel, 1956, p. 11) *contingent on H_0 being true*. Second, p is not the probability that the result is due to chance. This follows from the first, as well as from Falk & Greenbaum's (1995) point that α is a conditional probability. Specifically, the associated probability is the conditional probability, $p(\text{Data}|H_0)$.[3] Third, the complement of p (viz., $1 - .037 = .963$ in our example) is not informative as to the probability that the substantive hypothesis is true. Fourth, p is also not a measure of reliability or replicability of the results (Bakan, 1966). Fifth, researchers may be misled to believe that $p(H_0|\text{Data})$ is small simply because $p(\text{Data}|H_0)$ is small (Falk & Greenbaum, 1995).

1.7.   Limitations of NHSTP

The aforementioned discussion of p is instructive in the sense that it reveals what critics think researchers need, but fail, to know when they carry out statistical analyses (viz., the probability that the substantive hypothesis is true). Hence, what has been said about p effectively indicates what cannot be achieved with NHSTP. Some additional limitations of NHSTP at the practical level may now be noted.

At best, the outcome of NHSTP answers the question as to whether or not there is an effect. However, it is not informative about the magnitude of the effect. To many critics, this is the more important question. It is for this reason that researchers are urged to determine the size of the effect (J. Cohen, 1987, 1992a; Folger 1989; Harris, 1991; Rosenthal, 1984; Rosnow & Rosenthal, 1989).

As it is important for researchers to know the likelihood that their research will succeed, it seems necessary to know the probability of achieving statistical significance. However, this information is not available to users of NHSTP (J. Cohen, 1987, 1990). Hence, researchers are advised to consider statistical power in planning their research (J. Cohen, 1987, 1990, 1992a, 1992b, 1994). Moreoever, it has been shown that, at the mathematical level, whatever can be done with the analysis of variance (ANOVA) procedure (which is a NHSTP procedure) can be done better with linear regression analysis (J. Cohen & P. Cohen, 1975), a view endorsed by Oakes (1986). For example, one cannot make predictions on the basis of the ANOVA results. However, it is possible to do so on the basis of the coefficient of a linear regression analysis.

1.8.   Arbitrary Features in NHSTP

In addition to the arbitrary α level, another arbitrary feature of NHSTP is implied in Bakan's (1966) definition that the null hypothesis is the to-be-nullified hypothesis (see Section 1.3 above). Consider the example in Section 1.2 in a different context. Suppose that the research hypothesis of interest is $u_E - u_C = 0$. That is, the hypothesis of no difference is, in fact, the substantive hypothesis. The null hypothesis would be $u_E - u_C \neq 0$ under such circumstances (Meehl, 1967; Rozeboom, 1960). It appears as though the choice of the null hypothesis is a matter of convenience or the researcher's vested interests (Bakan, 1966). A third arbitrary feature in NHSTP is the size of the sample used. Two samples of 10 subjects each are used in the example in Section 1.2. However, before the advent of power analysis there was no objective way to determine what the correct sample size might be.

Recall from Section 1.3 Rozeboom's (1960) suggestion that, instead of predicting H_1: $u_E - u_C \neq 0$, a researcher may predict (a) H_1: $u_E - u_C = 5$, (b) H_1: $u_E - u_C = 10$, (c) H_1: $u_E - u_C = 15$ or any other value. Suppose a decision is made to test H_1: $u_E - u_C = 5$. The fourth arbitrary feature revealed by this example is that one out of many possible alternative hypotheses is being favoured at the expense of the others. The fifth arbitrary feature may be seen from the following statement:

> It is nonsensical, for example, to come to one conclusion with an associated probability of 0.048 and a radically different conclusion with a value of 0.052. And neither conclusion is safe if power and effect size have not been computed. (Oakes, 1986, p. 66)

[P1-1]

1.9.   Artefacts and Anomalies in NHSTP

In sum, some critics object to using NHSTP because of its various arbitrary features. These arbitrary features are objectionable because they may lead to artefacts or anomalies. Critics argue that statistical significance may be an artefact of the choice of the sample size. Significance is assured if a large enough sample is used. On the other hand, a non-significant effect may be the result of using too small a sample.

Anomalies may arise from the fact that statistical significance bears no relationship to the effect size or substantive importance. Specifically, a significant result may be a trivial one in practical terms. Alternatively, an important real-life effect may be ignored simply because it does not reach the arbitrarily chosen level of statistical significance. An example is Rosnow and Rosenthal's (1989) *ex post facto* analysis of a study of the efficacy of aspirin in reducing fatal myocardial infarction (MI) among MI patients. Their analysis showed that aspirin reduced the fatality by a half. Yet, the result was statistically insignificant.

It is generally agreed among critics that scientific knowledge is accumulated in an incremental fashion. However, NHSTP is a binary process. How is it possible to achieve the incremental growth of knowledge with a binary research procedure (Grant, 1962; Nunnally, 1960; Schmidt, 1992, 1996).

1.10.   Consequences of Using NHSTP

As has been noted in Section 1.5, to critics, NHSTP users ignore the importance of the Type II error. NHSTP users seem to take it for granted that a Type I error is more important than a Type II error. Critics do not accept this assumption. Moreover, this is not the only problematic consequence of using NHSTP. It is suggested that the common practice of using NHSTP is responsible for a simple-minded trial-and-error approach to experimentation, as witnessed by the following comment:

> The emphasis placed by significance tests upon the truth or falsehood of the null hypothesis . . . also encourages what one might term a manipulative mentality. . . . 'would it make any difference if I varied this factor?' (Oakes, 1986, p. 48)

[P1-2]

Furthermore, Gigerenzer (1993) notes that many researchers erroneously treat NHSTP as the *sine qua non* of scientific research. This may be responsible for the false impression of getting at the truth when NHSTP is being used. To some critics, using NHSTP 'actually obscures underlying regularities and processes in individual studies and in research literatures ...' (Schmidt, 1992, p. 1173). More important still, it has been noted that

> in practice there is a tendency to conflate *the substantive theory* with *the statistical hypothesis*, thereby illicitly conferring upon [the substantive theory] somewhat the same degree of support given by [the statistical hypothesis] by a successful refutation of the null hypothesis. (Meehl, 1967, p. 107; my emphasis)
>
> [P1-3]

This conflation may be the reason why some researchers are misled to such an extent that 'the publication of "significant" results tends to stop further investigation' (Bakan, 1966, reprinted in Badia, Haber & Runyon, 1970, p. 245). Perhaps the most severe consequence of using NHSTP may be the methodological paradox identified by Meehl (1967).

1.11. A Methodological Paradox

Meehl (1967) first notes that an experiment may be improved by reducing the probability of committing a Type II error while holding the probability of committing a Type I error constant. Meehl's (1967) point may be explicated as follows. The improvement in question is achieved by improving the design of the experiment. This improvement may take the form of (a) improving the implicative relation between the to-be-tested substantive hypothesis and the experimental hypothesis (see Section 3.2 below), and (b) choosing a more appropriate design for the experiment (see Section 4.5). Meehl (1967) also mentions increasing the size of the sample as a means of improving the experiment.

Suppose that such an improvement is carried out by a physicist, as well as by a psychologist. The physicist would obtain stronger support for a theory in physics, whereas the psychologist would obtain weaker support for a theory in psychology. This methodological paradox arises because the physicist does not rely on statistical significance, but the psychologist does (Meehl, 1978).

Consider first what the physicist does. The physicist may predict, on the basis of the to-be-tested theory, that $u_T = 5$ (i.e., a point-prediction). An estimate of u_T (i.e., u_E) is obtained on the basis of the sample mean. Whether or not there is experimental support for the theory depends on the difference between u_T and u_E. This comparison may be achieved by obtaining a 95% confidence-interval estimate based on u_E and the standard error of the mean (see Section 2.5.2). The theory is supported if u_T falls within the said interval.

A consequence of improving the experiment is the reduction of the standard error of the test statistic (viz., a smaller standard error of the mean in the present example). Consequently, the 95% confidence interval based on u_E is narrower the smaller the standard error of the mean is. That is, it is easier for the interval *to fail* to include u_T. This effectively means that a more stringent test is used to test the theory in physics when the experiment is improved. The physicist receives stronger support accordingly.

In using NHSTP to test a psychological theory, the psychologist begins with the null hypothesis which is always false. Meehl (1967) shows that, even if there is no relation between the to-be-tested theory and the experimental hypothesis, the long-run limit of the probability of obtaining a difference between the means of the experimental and control conditions is .5. He uses the following example (see also Oakes, 1986).

Suppose that there are two urns, an urn for theories and an urn for experimental designs. Further, suppose that one theory is randomly picked from the 'theory' urn, and that the expected outcome is $u_E - u_C > 0$ (i.e., H_1). One design is picked randomly from the 'situation' or 'design' urn. This means effectively that one condition is arbitrarily called the 'experimental condition' and the other, the 'control' condition. The main point is that the experimental manipulation is entirely independent of what the to-be-tested theory is.

Essential to the paradox is the assumption that the point-null hypothesis (H_0), $u_E - u_C = 0$, is always false. Hence, just by chance, half of the long-run differences between u_E and u_C will be positive (thereby, 'confirming' the theoretical expectation), and the other half will be negative. Any improvement made to the experiment will boost the probability of obtaining a positive difference above .5. Thus, it becomes easier to reject the null hypothesis. The support of the experimental hypothesis is an indirect one when NHSTP is used (viz., by rejecting the null hypothesis). It becomes easier to reject H_0 if there is a higher probability of rejecting it by chance to begin with. Consequently, the evidential support for H_1 becomes less convincing if it is obtained by an easier means.

1.12.   Summary and Conclusions

This brief review shows that virtually every aspect of NHSTP is subject to criticism. Specifically, there are problematic features in both the null and alternative hypotheses. Various arbitrary features of NHSTP may bring about artefacts and anomalies. Researchers cannot obtain certain important information from NHSTP. Moreover, the very act of using NHSTP may lead researchers astray.

It may have been noticed that some of the criticisms fall into more than one problem category (e.g., the difficulties with the concept *significance* and with the limited utility of NHSTP). Moreover, the criticisms raised by different critics may contradict one another. For example, to Meehl (1967), it is easier to reject the null hypothesis when the technical, non-statistical features of the research are improved. However, in Wilson and Miller's

(1964) view, more imprecise experiments (presumably as a result of unsatisfactory technicalities) are said to make it easier to reject the null hypothesis.

Be that as it may, the overall assessment of NHSTP is not encouraging. The puzzle is why many social scientists persist in using the process. It has been suggested that the continual reliance on NHSTP is a matter of inertia, conceptual confusion, submission to statistical authority, the lack of viable alternative tools and the positivistic 'demand for verification' (Oakes, 1986, p. 70; see also Falk & Greenbaum, 1995).

However, an alternative answer to why social scientists still use NHSTP is offered in this defence. It is argued that NHSTP is not as unsatisfactory as depicted by its critics. Moreover, it is possible to agree with the sentiments about empirical research expressed by NHSTP critics without abandoning NHSTP. After all, it is clear that the objective of the critics is to ensure that empirical research is conducted in a valid way. Suggestive of this possibility is critics' concern that NHSTP may be used incorrectly.

Strictly speaking, the criticisms in Section 1.10 concerning the consequences of using NHSTP can be discounted readily as criticisms of NHSTP itself. That someone misuses a tool does not necessarily speak ill of the tool. The problem may be due to the user. For example, the user may have misread the instruction accompanying the tool. Alternatively, the user may have been misled by the instruction provided by the tool manufacturer. None the less, it is helpful to consider which features of the research process in general, and of NHSTP in particular, may have misled empirical researchers.

It may also be seen that many of the criticisms of NHSTP are meta-theoretical or logical concerns. They are not statistical issues. For example, the substantive hypothesis in the example in Section 1.2 has a utilitarian overtone. It asks a 'Whether or not' question. Would it make any difference to the evaluation of NHSTP if the substantive hypothesis were one about why a certain phenomenon occurs? Related to this issue is the fact that some of the criticisms belong to domains radically different from statistics. For example, questions about the validity of a research procedure and questions about the real-life importance of the research results cannot (and should not) be treated as statistical issues.

These meta-theoretical questions may be appreciated only when the implicit assumptions underlying criticisms of NHSTP are made explicit. As an example, consider the assertion that NHSTP is wanting because merely knowing that the result is statistically significant is not useful to meta-analysis. To accept this criticism is to accept that meta-analysis is a valid procedure for corroborating theories. This assumption will be critically examined in Section 5.8 below.

Certain reasons for using NHSTP may become obvious only when some hitherto neglected distinctions are made. For example, some critics note that the cogency of their criticisms of NHSTP depends on the type of empirical research involved. Specifically, Meehl (1978) distinguishes between two kinds of experiment: (1) subject-variable experiments

(experiments in which only subject variables such as sex, ethnicity, age, etc. are used as independent variables), and (2) manipulated-variable experiments (experiments in which the independent variables are manipulated, such as method of teaching, amount of training, etc). The difference between the two kinds of variable is that, while the experimenter can assign subjects randomly to the two levels of method of teaching, such a manipulation is not possible in the case of subject variables (or 'assigned' variables in Kerlinger's, 1964, terms). The experimenter uses a subject variable to select subjects. Meehl (1967) makes it clear that his criticisms of NHSTP are more applicable to subject-variable experiments than to manipulated-variable experiments. This raises the following question:

Why should NHSTP be more problematic in the case of *subject-variable experiments* **than** *manipulated-variable experiments?*

[Q1-1]

At the same time, J. Cohen (1994) makes it clear that his comments about NHSTP apply only to experiments, not to true experiments. Similarly, Meehl (1990) distinguishes between non-experiments and experiments. Neither author actually explains the two respective distinctions. None the less, it seems reasonable to suggest that J. Cohen's (1994) true experiment is Meehl's (1967) experiment, and that J. Cohen's (1994) experiment is Meehl's (1967) non-experiment. 'Experiment' is used in subsequent chapters in the sense of J. Cohen's (1994) true experiment. The following question is crucial:

What renders NHSTP more satisfactory in an experiment than in a non-experiment?

[Q1-2]

[Q1-1] and [Q1-2] suggest that any discussion of NHSTP is incomplete, if not misleading, when the context of its application is not taken into account (i.e., whether it is used in an experiment or a non-experimental study). Another issue that has been neglected in criticisms of NHSTP so far is Meehl's (1967) distinction between the substantive theory (or substantive hypothesis) and the statistical hypothesis in [P1-3] in Section 1.10. It is not easy to appreciate this distinction with reference to the example in Section 1.2. The present defence of NHSTP will begin in the next chapter with the rationale of NHSTP.

Notes

1. Falk and Greenbaum (1995) find this manner of talking about Type I and Type II errors wanting for the following reason: the conditional nature of the two types of error may be overlooked if the descriptions are not read carefully. They have a good point. Be that as it may, the more exacting way of talking about these errors will be deferred till Section 2.2.4.

2. This used to be a difficulty when NHSTP users had to rely on using prepared tables at the back of a statistics textbook. However, it is no longer an issue with the advent of statistics software. The associated probability of a test statistic is now routinely provided by the software.

3. By 'data' is meant observations as extreme as, or more extreme than, the data in question.

2

The Null-hypothesis Significance-test Procedure (NHSTP)

The rationale of NHSTP is described, and the consequences of its hybridism are discussed. NHSTP is both a statistical decision and an inferential procedure. The 'H_0 is never true' criticism of NHSTP is not convincing because (a) H_0 is never used as a categorical proposition in NHSTP, and (b) H_0 is a hypothesis about the data-collection situation, not the substantive hypothesis or its logical complement. An examination of the nature and role of the sampling distribution of the test statistic serves (a) to raise a question about the graphical representation of NHSTP, (b) to make explicit the meaning of H_0 as the explanation of data in terms of chance influences, and (c) to evaluate some of the consequences of the hybridism of NHSTP.

2.1. Introduction

NHSTP owes its present form to Fisher (1959, 1960) and to the Neyman–Pearson contribution (the joint effort of J. Neyman and E.S. Pearson). A description of the similarities and differences between the two approaches is helpful for examining Gigerenzer's (1993) view that many of the problems with NHSTP are due to its hybridism. However, critics seem to have ignored the fact that the random sampling distribution of the test statistic determines (a) the probabilistic characteristics of NHSTP, and (b) the critical value of the test statistic, as well as its meaning. The graphical representation of NHSTP will be discussed with reference to the random sampling distribution of the test statistic.

2.2. Features of NHSTP

Table 1.1 gives an adequate description of the sequence of events implicated in testing a statistical hypothesis. However, some important features of NHSTP are not readily seen from the table. First, it is inevitable that one of two kinds of error is committed in NHSTP, namely, a Type I error or a Type II error. The second feature is that the choice of the α level is arbitrary.

Third, a disjunctive syllogism renders it possible to make a decision about H_1 via making the decision about H_0. The fourth feature is that its underlying mathematical tool provides NHSTP with its probabilistic characteristics. The mathematical tool is the random sampling distribution of the test statistic (the sampling distribution for short in subsequent discussion). The present discussion of the sampling distribution serves to provide answers for some difficulties with NHSTP identified by its critics.

2.2.1. *Some Conventional Features of NHSTP*

Three features in NHSTP are determined by convention. As has been noted, the first one is the choice of the α level. Two more conventional features should be mentioned in view of the contemporary concerns with the statistical power and effect size. Suppose that the aim is to estimate the size of the effect. To ensure that an appropriate sample size is used, it is necessary to determine the level of statistical power preferred (high, medium or low). On the other hand, if it is necessary to ensure that the statistical test is of sufficient power, a decision has to be made with regard to the expected effect size (large, medium or small). As an illustration of the conventional levels for the effect size, it is said that,

> for the test that $r = 0$, small, medium, and large [effect sizes] are, respectively, the population r_s .10, .30, and .50. For the test that two population means are equal, the [effect sizes], in the same order, are $d =$ .20, .50, and .80. (J. Cohen, 1992a, p. 99)

It is not the present intention to question the propriety or usefulness of choosing the required effect size or statistical power by a conventional rule. The issue is raised in view of the fact that some critics find fault with NHSTP because the α level is determined by convention (e.g., Glass et al., 1981). At the same time, they also recommend choosing the effect size by convention so as to establish statistical power, or selecting a conventional level of statistical power in order to ensure that the effect size so determined is real. It is not clear why the appeal to convention is acceptable for the effect size or statistical power, but not for statistical significance. It may be suggested that the actual issue goes beyond the fact that the choice of the α level is arbitrary. More will be said about this issue in Sections 2.8.1 and 5.2 below. Meanwhile, consider the critical value of the test statistic.

2.2.2. *A Conditional Syllogism and the Criterion Value*

In actual practice, the criterion value of $t_{(df=18)} = 1.734$ (or the critical t) for the .05 level mentioned in Table 1.1 is read from a table. The decision about statistical significance is made by comparing the calculated t (i.e., 2.05 in the example) with the critical t in a binary way to determine whether or not the calculated t equals, or exceeds, the critical t value. The outcome of this

Table 2.1 *Two conditional syllogisms (upper panel) and the disjunctive syllogism (lower panel) implicated in the null-hypothesis significance testing procedure (NHSTP)*

	Criterion exceeded	Criterion not exceeded
Major Premiss	*If* Calculated $t \geq$ (criterion = 1.734), *then* not H_0.	*If* Calculated $t <$ (criterion = 1.734), *then* H_0.
Minor Premiss*	$t \geq$ (criterion = 1.734) [e.g., Calculated $t = 2.05$]	$t <$ (criterion = 1.734) [e.g., Calculated $t = 1.56$]
Conclusion	Not H_0	H_0

	Statistical significance obtained
Major premiss:	H_1 or H_0
Minor premiss:	not H_0
Conclusion:	Therefore, H_1

*From Step 8 in Table 1.1

comparison (i.e., Step 8 in Table 1.1) determines which of the two conditional syllogisms shown in the upper panel of Table 2.1 to use.

The 'Criterion exceeded' syllogism is used when the criterion is exceeded. The 'Criterion not exceeded' syllogism is used when the criterion is not met. Furthermore, the outcome of the comparison between the calculated and critical t values serves as the minor premiss of the chosen syllogism (see the 'Minor Premiss' row in the upper panel of the table). In either case, this minor premiss is the antecedent of the conditional proposition that serves as the major premiss of the syllogism.

2.2.3. *A Disjunctive Syllogism*

In the event that criterion depicted in Table 2.1 is not obtained (i.e., the 'Criterion not exceeded' syllogism), the conclusion is drawn that H_0 is true. This syllogism shows that Rozeboom's (1960) reluctance to accept H_0 is unjustified when the result is insignificant for two reasons (see Frick, 1995, for further reasons). First, the minor premiss is chosen with reference to a well-defined criterion value (the critical t). Second, affirming the antecedent of the conditional proposition entails accepting the consequence of the major premiss (i.e., the *modus ponens* rule in deductive logic).

If, on the other hand, the criterion is exceeded, the conclusion of the 'Criterion exceeded' syllogism is that H_0 is not true. This leads to the disjunctive syllogism that is represented in the bottom panel of Table 2.1. As may be seen, the 'not H_0' conclusion of the 'Criterion exceeded' syllogism is used as the minor premiss of the disjunctive syllogism. It is by virtue of this disjunctive syllogism that researchers do not have to test H_1 directly. Instead, they accept H_1 by rejecting H_0.

It is necessary to iterate that the disjunctive syllogism depicted in the bottom panel of the table is used only when the 'Criterion exceeded' syllogism is applicable. Whether the conclusion is 'H_0' or 'not H_0', it is to be used in the next stage of the theory-corroboration process. This topic will be discussed in Sections 4.3 and 4.4 below. Meanwhile, it is necessary to note that the major premiss of the disjunctive syllogism in the lower panel of Table 2.1 is the disjunctive proposition, 'H_1 or H_0'. The validity of the syllogism depends on H_1 and H_0 being mutually exclusive and exhaustive. This major premiss is found to be problematic by some critics who argue that there are multiple H_1's (e.g., Neyman & Pearson, 1928; Rozeboom, 1960). Consideration of this issue at this juncture will lead to a digression from the present concern, which is to describe the features of NHSTP. Hence, the issue is deferred till Section 3.4. The next feature of NHSTP to examine is the two types of error found in NHSTP.

2.2.4.   Two Types of Error

The rationale of NHSTP is often represented graphically by two overlapping normal distributions, as depicted in the two panels of Figure 2.1. One distribution is based on H_0 (e.g., the 'Null' distribution on the left in both panels) and the other is based on H_1 (e.g., the 'Alternative' distribution on the right in both panels).[1]

It has been noted in Section 1.2 that two kinds of error are possible in NHSTP. It is necessary to emphasize the conditional nature of these two kinds of error (Falk & Greenbaum, 1995). A Type I error is made when the researcher decides to reject the null hypothesis, *given* that the null hypothesis is true (see Table 2.2). Its probability is set by the α level (see the area shaded with horizontal lines in both panels of Figure 2.1). A Type II error is committed when the researcher accepts (i.e., does not reject) the null hypothesis, *given* that the null hypothesis is false (as may be seen from Table 2.2). The probability of Type II error is represented by β (see the area shaded with vertical lines in both panels of Figure 2.1). As both types of error are errors contingent on another event, their probabilities are conditional probabilities.

The inverse relationship between α and β may also be illustrated with Figure 2.1. To begin with, it is necessary to note that the distance between the means of the two distributions is not known because there is no information about the mean of the H_1 distribution. That is, the probability of Type II error (i.e., β) is unknown. For this reason, the area shaded with vertical lines in both panels is actually ill-defined. None the less, it may be seen that α and β are inversely related, if the distance between the H_0 and H_1 distributions is held constant.

As may be seen from Figure 2.1, the probability of Type I error is changed

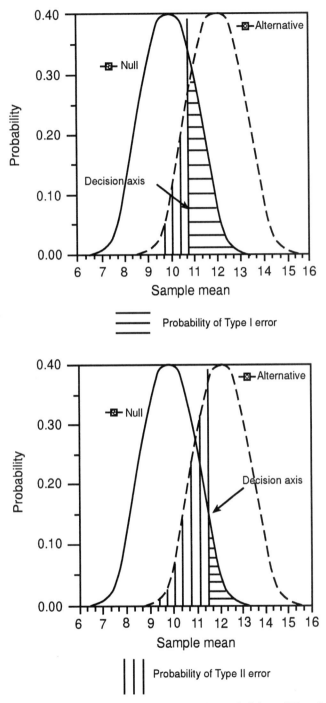

Figure 2.1   *The inverse relationship between the probability of Type I error
and the probability of Type II error*

Table 2.2   *Two types of error in NHSTP defined in terms of the truth value of the null hypothesis and the researcher's decision*

	State of affairs	
Decision	H_0 true	H_0 not true
Reject H_0	Type I error	Correct rejection of H_0
Accept H_0	Correct retention of H_0	Type II error

when the decision axis is moved. A larger probability of Type I error is shown in the left hand panel than in the right hand panel, as witnessed by the larger horizontally shaded area in the left hand panel. At the same time, the left hand panel represents a smaller probability of Type II error than the right hand panel (see the smaller vertically shaded area in the left hand panel).

The distance between the means of the two distributions is an index of the effect size (e.g., d1 in the left hand panel and d2 in the right hand panel of Figure 2.2). However, the distance is an unknown value because the mean of the H_1 distribution is not known. None the less, it is possible to illustrate that the probability of Type II error (i.e., the area shaded with vertical lines) is inversely related to the effect size with reference to the two one-tailed cases in Figure 2.2. The same α value is used in both panels (see the area shaded with horizontal lines). The effect size is smaller in the left hand than the right hand panel as witnessed by the fact that d1 is smaller than d2. At the same time, the probability of Type II error (i.e., the vertically shaded area) is larger in the left hand panel than in the right hand panel.

Another important numerical index is statistical power, defined as $(1 - \beta)$ (see the diagonally shaded area in both panels of Figure 2.2). As β is not known, neither is the statistical power known. That is, the diagonally shaded area in both panels is ill-defined. In actual practice, the determination of statistical power is said to depend on (a) the α level, (b) the effect size, and (c) the sample size (J. Cohen, 1987, 1992b; Kraemer & Thiemann, 1987). Moreover, an intimate relationship between statistical significance and statistical power has been suggested. Specifically, '[the] power of a statistical test is the probability that it will yield statistically significant results' (J. Cohen, 1987, p. 1). More will be said about this putative relationship in Chapter 6.

2.2.5.   *The One-tailed Test*

For tests involving one or two samples, a distinction is made between the one-tailed and two-tailed tests. The direction of the departure from H_0 is specified in the former case, but not the latter. Moreover, the null hypothesis of

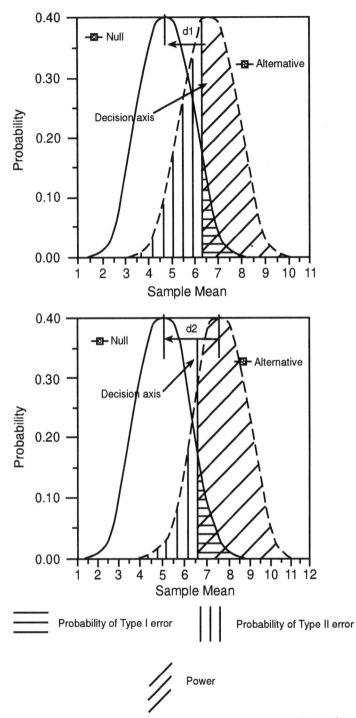

Figure 2.2   *The positive relationship between the effect size* (d1 *in the upper panel and* d2 *in the lower panel*) *and the power of the test*

the one-tailed test is not a simple proposition. Instead, it is a complex of two propositions: 'H_0: $u_E - u_C \leq 0$' or 'H_0: $u_E - u_C \geq 0$'.

That the directional null hypothesis is made up of two propositions is not a complication at all. This may be illustrated with 'H_0: $u_E - u_C \leq 0$' as an example (i.e., H_1: $u_E - u_C > 0$) with reference to Figure 2.3. The two propositions making up 'H_0: $u_E - u_C \leq 0$' are '$H_{0'}$: $u_E - u_C < 0$' (viz., the left or 'Negative-null' distribution in the figure) and 'H_0: $u_E - u_C = 0$' (i.e., the right or 'Point-null' distribution in the figure). Note that the horizontal axis of Figure 2.3 represents the value of the *t* statistic. (See Section 2.7.1 for an explanation of this change from Figures 2.1 and 2.2 in how the horizontal axis is labelled, i.e., from 'Sample mean' to 'Value of *t* statistic'.)

The decision axes of the negative-null ($H_{0'}$) and the point-null (H_0) are Decision Axis DA < 0 and Decision Axis DA $= 0$, respectively. As may be seen, DA < 0 is to the left of DA $= 0$ on the continuum of *t* values. In other words, DA < 0 is *less extreme* than DA $= 0$ with reference to the *positive end* of the continuum of *t* values. Consequently, whenever a calculated *t* falls to the right of DA $= 0$ (i.e., the region of rejection of the point-null hypothesis shaded horizontally), it automatically falls to the right of DA < 0. That is to say, whatever is rejected in terms of DA $= 0$ is necessarily rejected in terms of DA < 0. A similar argument applies to the case of 'H_0: $u_E - u_C \geq 0$'. For this reason, 'H_0: $u_E - u_C \leq 0$' or 'H_0: $u_E - u_C \geq 0$' is justifiably used as though it is 'H_0: $u_E - u_C = 0$'.

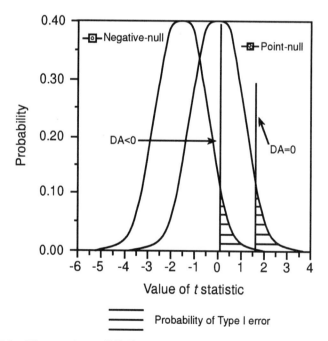

Figure 2.3   *The negative-null (left) and point-null (right) distributions for the positive directional alternative hypothesis*

2.3. The Hybridism of NHSTP

NHSTP has been given two characterizations. The first is that it is 'the framework of Neyman–Pearson orthodoxy' (Oakes, 1986, p. vii). Supportive of this description are the boldface entries in the 'Neyman–Pearson' column in Table 2.3. The Neyman–Pearson features found currently in NHSTP are as follows. First, H_0 is the hypothesis of zero difference (Row 6). Second, the α level must be chosen before data analysis, and it must remain unchanged (Row 9). Third, there are two types of error (viz., Types I and II; see Row 10). Fourth, separate distributions are used to represent graphically H_0 and H_1 (Row 11).

Table 2.3 *The similarities and differences between the Neyman–Pearson and the Fisherian approaches to inferential statistics*

Feature	Neyman–Pearson	Fisher
1 Statistical procedure	***Indistinguishable from the scientific procedure***	
2 Objective of NHSTP	To determine the acceptability of H_0	To determine the truth of a hypothesis
3 Nature of NHSTP	**Decision making**	*An inferential (inductive) procedure*
4 Random sampling distribution	The actual mechanism by which the data were obtained – to repeat the experiment many times	A framework for assessing data (a logical basis for calculating the associated probability, p)
5 Probability of interest	The inverse probability, $p(H|D)$	*The associated probability, $p(D|H_0)$*
6 H_0	**The hypothesis of zero difference**	The to-be-nullified hypothesis
7 H_1	***The substantive hypothesis itself***	
8 Disjunctive syllogism	Not applicable because of multiple H_1's	*Applicable; exclusive OR relationship between H_0 and H_1*
9 The α Level	**The α must be chosen before data analysis**	Both associated probability, p, and conventional α
10 Error	**Types I and II errors**	Type I error only
11 Graphical representation	**Separate distributions for H_0 and H_1**	One distribution for H_0 only
12 Statistical power	The power of a test was anticipated	The power of a test was not anticipated
13 Research objective	Qualitative control issues	Scientific inference

Notes
Italicized boldface entries = features in NHSTP found in both the Neyman–Pearson and Fisherian approaches
Entries in Roman font = features not emphasized in contemporary discussion of NHSTP
Entries in italics = features in NHSTP originating from the Fisherian approach
Boldface entries = features in NHSTP originating from the Neyman–Pearson approach

The second characterization of NHSTP is that it is a hybrid of the Fisherian and the Neyman–Pearson approaches (Gigerenzer, 1993). That is to say, in addition to the Neyman–Pearson features, other features of NHSTP owe their origin to the Fisherian approach. The Fisherian features found in the contemporary approach to NHSTP are the entries in *italics* in the 'Fisher' column in Table 2.3, namely, (a) NHSTP is an inferential procedure (specifically, an inductive procedure to some critics of NHSTP; see Row 3), (b) the inference is based on a conditional probability, the (associated) probability of obtaining the data, given that H_0 is true (i.e., $p(\text{Data}|H_0)$) (See Row 5), and (c) H_0 and H_1 are mutually exclusive and exhaustive (Row 8).

As the Fisherian and Neyman–Pearson approaches have something in common, it is not surprising that two meta-theoretical assumptions underlying NHSTP may be traced to both schools of thought. These assumptions are the *italicized* **boldface** entries in Table 2.3. The first one is the meta-theoretical assumption that to carry out NHSTP is to carry out a scientific procedure (Row 1). The second assumption is that H_1 is the substantive hypothesis itself (Row 7).

The rest of the entries in Roman font in Table 2.3 are features in the Fisherian or Neyman–Pearson approach that are not often dealt with in introductory textbooks. Of interest is the Neyman–Pearson view that there are multiple H_1's (see Row 8 of Table 2.3). This feature is not mentioned in introductory textbooks. However, it is one of the features emphasized by critics in contemporary discussion of NHSTP. Some of these features indicate the Fisherian or Neyman–Pearson view of what scientific research is. For example, to J. Neyman and E.S. Pearson (1928), the purpose of conducting scientific research was to ascertain the inverse probability (see Row 5 of Table 2.3). A good account of the Fisherian assumptions about scientific research may be found in Brenner-Golomb (1993).

2.4. Consequences of the Hybridism

Strictly speaking, that NHSTP owes its origin to two sources does not necessarily mean that it is problematic. However, the hybridism is an issue because NHSTP is said to be 'inconsistent from both perspectives and burdened with conceptual confusion' (Gigerenzer, 1993, p. 324). It is further suggested by Gigerenzer (1993) that the inconsistency and confusion lead researchers, journal editors and especially textbook writers to display compulsive and ritualistic behaviour, as well as dogmatism, when they deal with NHSTP.

2.4.1. *A Psychological Account of Statistical Reasoning*

Gigerenzer (1993) argues that the hybridism of NHSTP leads to psychological tensions on the part of NHSTP users. He identifies three psychological components. First, the features originating from the Neyman–Pearson approach collectively make up the superego of statistical reasoning. These superego features serve a regulatory function. They prescribe what should be done, and hence are responsible for the compulsive and ritualistic behaviour of NHSTP users (e.g., setting the α level to .05 before data analysis).

Second, the ego is made up of the Fisherian features. They render it possible for the researchers to '[get] things done in the laboratory and [get] papers published' (Gigerenzer, 1993, p. 324), even at the risk of violating some of the Neyman–Pearson prescriptions. An example is testing H_1 via testing H_0 by assuming a mutually exclusive and exhaustive relationship between H_0 and H_1. However, to critics of NHSTP, there are multiple H_1's (e.g., Rozeboom, 1960; and see Section 1.3 above). NHSTP users suffer from tensions because of the conflicts between what the Neyman–Pearson features allow and what the Fisherian features permit.

The third Freudian component is the Bayesian wish to assign probabilities to theories on the basis of research data. This is the Neyman–Pearson's preference for the inverse probability, $p(\text{Hypothesis}|\text{Data})$. To Gigerenzer (1993), this preference for the inverse probability behaves like the Freudian id because it is censored and suppressed by the Fisherian subscription to the associated probability, $p(\text{Data}|\text{Hypothesis})$. None the less, the hidden Bayesian wish for using the inverse probability reflects the researcher's wish to have '(some) direct measure of the validity of the hypothesis under question – quantitatively or qualitatively . . .' (Gigerenzer, 1993, p. 325), a view recently endorsed by J. Cohen (1994), as well as by Falk and Green-baum (1995).

In sum, Gigerenzer (1993) suggests that the conflicts among these three psychological components give rise to distorted statistical intuitions and conceptual confusion. His account of the psychodynamics of NHSTP users is interesting and may well be true when NHSTP is misunderstood. In any case, a psychological account about NHSTP users is not an explication of what renders a statistical procedure unsatisfactory as a research tool. Nor is the psychological account informative as to what a satisfactory statistical procedure should be like. None the less, a positive reading of Gigerenzer's (1993) Freudian analogy is that NHSTP users and critics may actually have diverse views of what using NHSTP implicates because they entertain different meta-theoretical, non-statistical assumptions about the research process. Consequently, people on opposite sides of various disputes about NHSTP may actually be talking at cross purposes.

As an example, consider the relationship between H_0 and H_1. They are used as two mutually exclusive and exhaustive alternatives in the lower panel of Table 2.1. However, this relationship holds only if H_1 is stated in the quantitatively ill-defined manner, such as (i) H_1: $u_E - u_C > 0$; (ii) H_1: $u_E - u_C < 0$, or (iii) H_1: $u_E - u_C \neq 0$. It does not hold if H_1 is stated as a point-prediction, such as H_1: $u_E - u_C = 5$. Some critics, in fact, do find the qualitative nature of H_1 insufficient (e.g., Meehl, 1967; Oakes, 1986).

However, the important point is that the actual bone of contention is not a statistical issue. As may be seen from Sections 3.3 and 3.4 below, the dispute about the relationship between H_0 and H_1 is one about what H_0 and H_1 represent. That is to say, some disputes about NHSTP owe their origin to issues beyond statistics. Be that as it may, it is important to note that H_1 is implicitly assumed to be an integral part of NHSTP in both the Fisherian and the Neyman–Pearson approaches. This assumption is by no means a self-evident one, a point that will be developed in Sections 3.2.4 and 3.4 below.

2.4.2.   *The Putative Inconsistencies in NHSTP*

Suppose that Psychologist P subscribes to the Fisherian associated probability of the test statistic when using NHSTP (i.e., $p(D|H_0)$. Psychologist P entertains effectively an assumption inconsistent with the Neyman–Pearson preference for the inverse probability (see Row 5 in Table 2.3). At the same time, in choosing the α level before data analysis, and in considering both Type I and Type II errors, Psychologist P is doing something contrary to Fisher's position. In other words, Psychologist P is inconsistent with both the Fisherian and the Neyman–Pearson perspectives, a point made by Gigerenzer (1993).

However, the inconsistency thus identified says simply that NHSTP, as it is currently carried out, is neither purely Fisherian nor purely in the mould of the Neyman–Pearson approach. As such, the hybridism is not necessarily problematic. It is detrimental to research rigour only if it is established that either the Fisherian or the Neyman–Pearson approach has to be adopted in its entirety. Moreover, it has to be assumed that either the Fisherian or the Neyman–Pearson treatment of NHSTP is adequate by itself for the task. At the same time, the composite of NHSTP identified in the present discussion (viz., the Table 2.3 entries in **boldface** or *italics* or ***both***) is only one of many possible composites. For example, some researchers treat NHSTP explicitly as a decision-making, not an inferential, procedure (Tukey, 1960), despite the fact that NHSTP is commonly characterized as an inferential procedure (see Row 3 in Table 2.3).

The more relevant issue is whether or not NHSTP, as it is currently practised, is adequate for the task. This consideration involves examining three separate issues. The first one is the mathematical basis of NHSTP, a topic to be taken up in Section 2.5 below. The second issue is about the rationale of empirical research and the role of statistical analysis in empirical research; this will be discussed in Chapters 3 and 4. The third issue is the role of the researcher's subjective belief in empirical research. This will be dealt with when Bayesianism is considered in Chapter 7.

2.5.   The Mathematical Foundation of NHSTP

Although the mathematical foundation of NHSTP is the sampling distribution of the test statistic, the sampling distribution is seldom taken into account in criticisms of NHSTP. However, issues about the nature of H_0, the associated probability (p), the meaning of the α level and the like may become clearer if the nature of the sampling distribution and its role in NHSTP are made explicit. For this reason, the present discussion starts from the very beginning, even at the risk of stating the obvious.

2.5.1. Sampling Distribution

A good place to start the discussion of sampling distribution is Hays's (1994) definition:

> A sampling distribution is a theoretical probability distribution that shows the relation between the possible values of a given statistic and the probability (density) associated with each value, for all possible samples of size n drawn from a particular population. (Hays, 1994, p. 206; n is used to represent the sample size instead of the capital N in the original)

[D2-1]

Definition [D2-1] may be illustrated with a hypothetical population made up of seven scores, namely, 1, 2, 3, 4, 5, 6 and 7. Suppose that samples of Size 2 (i.e., $n = 2$) are to be drawn randomly from this population with replacement. Table 2.4 shows all possible samples of Size 2 drawn randomly, with replacement, from this population. It is sampling with replacement when a chosen element is returned to the population before the next element is selected. The sampling procedure is a random one when every one of the possible samples has an equal chance of being selected.

Further, suppose that the statistic of interest is the mean of a sample of two scores. Among the 49 possible samples of Size 2, there are 13 possible values of the mean (see Column 1 of Table 2.5) which occur with different frequencies, as shown in Column 2 of Table 2.5. Columns 1 and 2 of the table jointly form the frequency distribution of all possible sample means from the said population.

The frequencies in Column 2 of Table 2.5 can be converted into their respective relative frequencies (viz., probabilities), as in Column 3 of the table. In other words, an entry in Column 3 is the probability of the mean (in Column 1) in the same row. Columns 1 and 3 jointly form the sampling distribution of means of samples of Size 2 as defined in [D2-1] if the sampling procedure is carried out an infinite number of times. If the means in

Table 2.4   *All possible samples of Size 2 randomly chosen with replacement from Population P consisting of Scores 1, 2, 3, 4, 5, 6 and 7*

1,1	2,1	3,1	4,1	5,1	6,1	7,1
1,2	2,2	3,2	4,2	5,2	6,2	7,2
1,3	2,3	3,3	4,3	5,3	6,3	7,3
1,4	2,4	3,4	4,4	5,4	6,4	7,4
1,5	2,5	3,5	4,5	5,5	6,5	7,5
1,6	2,6	3,6	4,6	5,6	6,6	7,6
1,7	2,7	3,7	4,7	5,7	6,7	7,7

Table 2.5   *The frequency distribution of all possible values of the mean of samples of Size 2 From Population P consisting of Scores 1, 2, 3, 4, 5, 6 and 7*

1 Sample mean	2 Frequency of mean	3 **Probability** **of mean**	4 z of mean	5 Area beyond z
7	1	.0204	2.12	0.0170
6.5	2	.0408	1.76	0.0392
6	3	.0612	1.41	0.0793
5.5	4	.0816	1.06	0.1446
5	5	.1020	0.71	0.2389
4.5	6	.1224	0.35	0.3632
4	7	.1428	0.00	0.5000
3.5	6	.1224	−0.35	0.3632
3	5	.1020	−0.71	0.2389
2.5	4	.0816	−1.06	0.1446
2	3	.0612	−1.41	0.0793
1.5	2	.0408	−1.76	0.0392
1	1	.0204	−2.12	0.0170

Column 1 are converted into their respective z scores (as in Column 4), then Columns 4 and 3 together represent the sampling distribution of means in standard scores.

2.5.2.   *Sampling Distribution of Means*

Several things may be said about the sampling distribution of means. First, it is the distribution contingent on chance by virtue of the fact that the random selection procedure is used. Second, it does *not* follow from Table 2.5 that the entries in Columns 1 and 3 are obtained in every block of 49 attempts (or trials) to randomly select two units with replacement. Nor does it mean that, given a particular number of trials (say, 130), each of the possible means occurs equally often (viz., 10 in the present case). What it means is that entries in Columns 1 and 3 are obtained, *in the long run*, if random sampling is carried out. Specifically, the 'in the long run' characterization means an infinite number of trials in which (a) a sample of Size 2 is randomly selected with replacement, (b) the mean of the chosen sample is calculated, (c) the selected units are returned to the population, and (d) Steps (a) and (b) are repeated.

This discussion of the sampling distribution is relevant to the replication fallacy identified by Gigerenzer (1993). The thesis of the fallacy is that the α level is incorrectly interpreted to refer to the percentage of exact replications of a study that will give significant results.[2] However, the present description of the nature of the sampling distribution shows that nothing inherent in NHSTP encourages this wrong interpretation. In other words, the fact that

NHSTP is misunderstood, or misrepresented (particularly by its critics), does not necessarily mean that it is problematic.

Third, this long-run probability density function is, in fact, a mathematical construct based on the Central Limit Theorem. It is for this reason that the sampling distribution of means (or any sampling distribution used in inferential statistics) is a theoretical distribution. It should also be emphasized that this theoretical basis is a mathematical one. It is independent of any substantive theory in the content area.

Fourth, the sampling distribution of means approximates a normal distribution, regardless of the shape of the original population. The approximation to the normal distribution is better the larger the sample size. The mean of means (i.e., $u_{\bar{x}}$) is equal to the population mean, and the standard error of the mean (viz., $\sigma_{\bar{x}}$) is $\sigma/\sqrt{n}$, or $s/\sqrt{n}$ if σ is not known. It is for this reason that the sampling distribution of means is said to be derived from the population mean (i.e., u) and the population standard deviation (viz., σ) or their corresponding sample statistics. Fifth, representing the sampling distribution of means by the z or t distribution makes it possible to have a procedure which enjoys generality (see Siegel, 1956).

2.5.3. *Sampling Distribution of Differences between Two Means*

The idea of the sampling distribution of means can be extended to the concept of the sampling distribution of differences between two means of samples of sizes n_1 and n_2 (called the sampling distribution of differences for short in subsequent discussion). This may be illustrated with two populations of identical composition, as shown in the left hand panel of Table 2.6[3] Suppose that the following sequence of events is carried out:

(1) A sample of 15 is selected randomly, with replacement, from each of the two populations.
(2) The means of the two samples are calculated (e.g., $\overline{X}_E$ and $\overline{X}_C$ are 4.73 and 5.00, respectively, on Trial 1 in the right hand panel of Table 2.6).
(3) The difference between the two sample means is determined (viz., -0.27 on Trial 1).
(4) The units of the two samples are returned to their respective populations.
(5) Steps (1)–(4) are repeated 25 times.

Each of the 25 entries in the $(\overline{X}_E - \overline{X}_C)$ column in Table 2.6 is the difference between two sample means. The mean of these 25 differences is the mean difference, whose value is 0.064. The difference between the two population means is 0. If the frequency distribution of these 25 differences is tabulated, its standard deviation (i.e., the standard error of the difference) is 0.298.

Suppose that Step (5) is carried out not just 25 times, but an infinite number of times. There will be an infinite number of differences between two means. A frequency distribution for this infinite number of differences

Table 2.6 *A 25-trial approximation to the sampling distribution of the differences between the means of two independent samples of Sizes 15 and 15 (panel (b)) from two populations of the same composition (panel (a))*

(a) The composition of two identical populations		(b) 25 sample means and their differences			
Score	Frequency	Trial	$\overline{X}_E$ $(n = 15)$	$\overline{X}_C$ $(n = 15)$	$(\overline{X}_E - \overline{X}_C)$
1	1	1	4.73	5.00	−0.27
2	12	2	5.20	4.67	0.53
3	36	3	4.67	5.00	−0.33
4	412	4	4.73	4.73	0.00
5	669	5	4.27	4.73	−0.47
6	128	6	4.40	4.87	−0.47
7	65	7	4.87	5.07	−0.20
8	5	8	4.80	4.73	0.07
		9	4.73	4.87	−0.13
$N_1 = N_2$	1328	10	4.93	4.87	0.07
$u_1 = u_2$	4.81	11	5.20	4.60	0.60
$\sigma_1 = \sigma_2$	0.947	12	5.00	4.40	0.60
		13	5.07	4.67	0.40
		14	4.93	4.60	0.33
		15	4.60	4.67	−0.07
		16	5.00	4.87	0.13
		17	4.87	4.87	0.00
		18	4.73	4.67	0.07
		19	4.87	4.80	0.07
		20	4.80	4.67	0.13
		21	4.80	4.93	−0.13
		22	4.87	4.53	0.33
		23	4.87	4.73	0.13
		24	4.73	4.80	−0.07
		25	4.67	4.40	0.27

Mean difference = 0.064
Standard error of differences = 0.298

may be formed as Columns 1 and 2 of Table 2.5 are constructed. By the same token, a probability distribution of all the possible values of the difference between two means may be formed after the fashion of Columns 1 and 3 of Table 2.5. The outcome of this exercise is the sampling distribution of differences. Furthermore, the probability distribution so formed may be expressed in terms of standard scores (i.e., the distribution formed as Columns 3 and 4 of Table 2.5 are formed).

As is the case with the sampling distribution of means, it is not necessary to carry out the actual sampling and calculation repeatedly. That is, the sampling distribution of differences is also a theoretical probability density function. Moreover, it is likewise a theoretical distribution contingent on chance, in the sense that its validity depends on the assumption that the difference between $\overline{X}_E$ and $\overline{X}_C$ on any trial is the result of chance influences (i.e., the sequence of events depicted in Steps 1–5). This between-groups difference within a trial is also responsible for the inter-trial variation. The sampling distribution is normal in shape. Moreover, its mean (i.e., the mean difference) and standard deviation (viz., the standard error of the difference) can be calculated from the corresponding parameters, or the sample estimates of the parameters, of the two populations.

2.5.4. Standardized Representations of Sampling Distributions

Consider again the means of two independent samples on Trial 1 in the right hand panel of Table 2.6 as an example. The difference between $\overline{X}_E$ and $\overline{X}_C$ is -0.27. Suppose that the two respective standard deviations of the two samples are 0.7 and 1.0. It follows that the standard error of the difference for this independent-sample t test is 0.32 with df $= (n_1 + n_2 - 2) = 28$ (see, e.g., Kirk, 1984, pp. 287–8). Hence, the t statistic for this difference of -0.27 is $-0.270/0.32 = -0.84$. That is to say, the t statistic is simply the difference score (viz., -0.27) expressed in the appropriate standard error of the difference units (i.e., 0.32 in the case of Trial 1 in Table 2.6).[4]

The general point illustrated by this example is that, in using the t distribution, the researcher is appealing to the sampling distribution of $(\overline{X}_E - \overline{X}_C)$ expressed in terms of the appropriate standard error units. It is in this sense that Siegel (1956) treats theoretical distributions such as t, z, χ^2, F and the like as sampling distributions.

The theoretical nature of the sampling distribution of the test statistic renders it possible to answer the following questions: (a) What is the probability of getting a test statistic as extreme as, or more extreme than, a criterion value (e.g., $t_{(df=18)} \geq 1.734$)? (b) What is the probability of getting a test statistic within a particular interval? (c) Which pair of means (or differences) includes 95% of all possible means (or differences)? That is to say, it is the appeal to the appropriate sampling distribution of the test statistic that gives NHSTP its probabilistic character. Moreover, this appeal is justified on the grounds of the data-collection procedure, not the substantive hypothesis.

This appeal to the sampling distribution also confers meanings to concepts important to NHSTP, namely: (a) H_0, (b) Type I error, (c) the critical values of t, χ^2, F and the like, (d) the associated probability, p, and (d) the α level. The sampling distribution is also the bridge between a sample statistic and its corresponding population parameter. What has been neglected hitherto is that the probabilistic character of NHSTP is a conditional one. This last issue leads to the consideration of the nature of H_0.

2.6. The Nature of H_0

It has been mentioned that H_0 is often defined as the hypothesis of no difference. Moreover, the following proposition is an important criticism of NHSTP (see Section 1.3 above).

The null hypothesis, H_0: $u_E = u_C$, is never true.   [P2-1]

Of interest to the present discussion is that H_0 is assumed to be used as a categorical proposition in NHSTP in [P2-1], not as a component of a conditional proposition. This has an important impact on the validity of the criticism.

2.6.1. *Proposition-Categorical versus Conditional*

A distinction needs to be made between a categorical and a conditional proposition. A *categorical proposition* is one whose truth value is not contingent on its relation to another proposition. An example is the following proposition, which is a false one:

Human beings can fly with their hands.   [P2-2]

A *conditional proposition* is one formed by linking two propositions with the logical operator, 'If ... then ...' The proposition headed by 'If' is the antecedent, and the proposition after 'then' is the consequent of the conditional proposition. The truth value of a conditional proposition is determined by the relationship between its antecedent and consequent (Copi, 1965, 1982). An example is Proposition [P2-3], which is a true one even though both its antecedent and consequent are false:

***If* human beings have feathers, *then* human beings can fly with their hands.** [P2-3]

The impact of the difference between a categorical and a conditional proposition on the criticism of H_0 may be seen by considering Propositions [P2-2] and [P2-3]. The false proposition, [P2-2], is the consequent of [P2-3]. Despite this fact, [P2-3] is a true proposition. This is the case because a conditional proposition is false only if its antecedent is true while its consequent is false. That both the antecedent and consequent of [P2-3] are false renders [P2-3] true. The general point is that the truth value of a conditional proposition is not determined solely by the truth value, its consequent or its antecedent.

2.6.2. *H_0, the Data-collection Procedure and Two Conditional Propositions*

Suppose that the null hypothesis were never true. Its role in NHSTP is not thereby suspect if the null hypothesis is never used as a categorical proposition. In actual fact, it can be shown that H_0 is never used as a categorical proposition in NHSTP. Specifically, H_0 appears twice every time NHSTP is carried out, once as the consequent of a conditional proposition (see [P2-4] below), and once as the antecedent of a second conditional proposition (see [P2-5] below). This state of affairs may be explicated by considering again the means, $\overline{X}_E$ and $\overline{X}_C$, of the two independent samples on Trial 1 in the right hand panel of Table 2.6.

Suppose that the independent variable is Method of Teaching, whose two levels are New Teaching Method (E) and Old Teaching Method (C). $\overline{X}_E$ and $\overline{X}_C$ are the means of (E) and (C), respectively; and it is customary to state, 'H_0: $u_E = u_C$'. However, this customary way of saying things gives a misleading view of the role of the null hypothesis in NHSTP. It is misleading because the relationship between H_0 and the data-collection procedure is not properly represented when H_0 is used as a categorical proposition in discussing NHSTP.

It is important to supplement the present example with some additional information about the data-collection condition. Specifically, assume that (a) 30 individuals are randomly selected from the population of interest (e.g., all high school students from a particular school district), and (b) the 30 individuals are randomly split into two groups of 15 (Group E and Group C) before being taught statistics.

A critical feature of the research manipulation is to do one thing to Group E and something else (or nothing) to Group C. At the level of statistical discourse, Group E is assumed to be a sample of the hypothetical population made up of all individuals who are being taught with the new method. By the same token, Group C is a sample of another hypothetical population made up of all individuals being taught with the old method. In short, the two groups may be treated as two samples coming from two respective populations, namely, the E-method and C-method populations. The research manipulation, if efficacious, renders different two initially identical theoretical populations. The two theoretical populations remain identical if the research manipulation is not efficacious (see Frick, 1995).

It is also important to recall that the two groups come originally from the same source. Hence, the two theoretical populations have the same mean before the research manipulation is carried out (i.e., $u_E = u_C$). Moreover, the two populations are comparable in ways relevant to the research question, a point to be developed in Section 4.5. Allowing for two theoretical populations is a more satisfactory characterization of the NHSTP situation than allowing for only one population, as in the assertion, '[the] null hypothesis of no effect in **the population** is, actually, the hypothesis of chance . . .' (Falk & Greenbaum, 1995, p. 88; emphasis in boldface added). However, in characterizing the null hypothesis as the 'hypothesis of chance', Falk and Greenbaum (1995) make explicit that the null hypothesis is not a substantive hypothesis or its complement. That is, NHSTP is not about the substantive hypothesis.

Whether or not the two theoretical populations are identical in terms of the parameter of interest (viz., u in the present example) at the completion of the research is assumed to be dependent on whether or not the two methods of teaching differ in efficacy. This, of course, is the empirical issue to be settled. That is, if a difference between Groups E and C is observed (as samples of two respective populations), it is assumed that the difference is due to either (a) the difference in efficacy between the two methods of teaching (as stipulated by the to-be-tested hypothesis), or (b) chance influences such as the result of the random selection and assignment (to the two conditions) of individuals.

In the event that the research manipulation is not efficacious,[5] chance influences may still produce a numerically non-zero difference between the two samples. However, the majority of the non-zero differences are very close to zero. Hence, if the research manipulation is not effective, it is possible and correct to say that '$u_E = u_C$' is true. The implicative relationship between the research manipulation and the two theoretical populations is represented in the form of the following conditional proposition:

***If* the research manipulation is not efficacious (i.e., only chance influences are assumed), *then* H$_0$.**

[P2-4]

In other words, H$_0$ serves as the necessary condition for the acceptance of the antecedent of the conditional proposition (viz., chance influences as the explanation of the research outcome). Put another way, H$_0$ is not used as a categorical proposition about the to-be-studied phenomenon. Instead, it is about the data-collection procedure. Moreover, it is contingent on assuming chance influences as the only source of influence, a point emphasized by Falk and Greenbaum (1995).

The assertion '$u_E = u_C$' is about two theoretical populations in the context of a specific data-collection situation. However, the data are from two samples. It is, hence, necessary to have a means to connect the two sets of sample data to their respective populations. Moreover, it is also important to have a rational basis for making assertions about the relationship between the two populations on the basis of the observed relationship between their respective sets of sample data. The tool required for these tasks is the sampling distribution of differences. Specifically, H$_0$ (i.e., $u_E = u_C$) stipulates using the sampling distribution of differences whose mean difference is zero. This implicative relationship between H$_0$ and the sampling distribution of differences is represented by a second conditional proposition:

***If* H$_0$, *then* the mean difference of the sampling distribution of differences is zero.**
[P2-5]

That is to say, the second occurrence of H$_0$ in NHSTP is also not in the form of a categorical proposition. Instead, it is the antecedent of a conditional proposition. It serves as the sufficient condition for choosing a particular sampling distribution.

The roles of H$_0$ in the two conditional propositions, [P2-4] and [P2-5], have now been discussed. It may be suggested that the 'The null hypothesis, H$_0$: $u_E = u_C$ is never true' criticism is misleading. It is not unreasonable to assume $u_E = u_C$ in the correct research context.[6] (See Section 3.5 and Section 4.5 for further discussions.)

2.7.   The Probabilistic Nature of NHSTP

Gigerenzer (1993) has given many examples of incorrect statements in statistics textbooks and journals made about NHSTP in general, and about statistical significance in particular (so do Falk & Greenbaum, 1995). While his observation is cogent, it cannot (and should not) be considered as an indication that something is problematic with NHSTP. Nor do Gigerenzer's (1993) or Falk & Greenbaum's (1995) exercises show how a better understanding of NHSTP may be achieved. None the less, the examples make it clear that it is imperative to use the NHSTP concepts more carefully.

To iterate, being the antecedent of [P2-5], H_0 prescribes the mean of the sampling distribution of differences. As the sampling distribution is a probability density function, H_0 determines effectively the probabilistic nature of NHSTP. The majority of incorrect statements identified by Gigerenzer (1993) may be avoided if it is kept in mind that many concepts in NHSTP owe their meanings to the sampling distribution of the test statistic, regardless of whether the researcher is using the z distribution, t distribution, F distribution, or any other distribution found in statistics textbooks. Moreover, the choice of the sampling distribution is contingent on assuming chance influences as the explanation in a particular context. It is equally important to foreshadow the thesis to be developed in Chapters 3 and 4 that NHSTP is not the theory-corroboration procedure.

2.7.1.   Tabular and Graphical Representations of NHSTP

Table 2.7 gives the one-tailed and two-tailed critical values of $t_{(df=18)}$. This table is used in conjunction with Figure 2.4 in the present discussion. The 'null' distribution in any panel of Figure 2.4 is the sampling distribution of differences chosen on the assumption that chance is the explanation of the research data. The top two panels of Figure 2.4 are used in one-tailed cases. The top left panel is used when the expected difference between u_E and u_C is negative. In such an event, all entries in the last row of Table 2.7 are treated as negative numbers, (viz., -1.734, -2.101, -2.552, and -2.878). The top right panel is used when the expected difference is positive. The entries in the last row of Table 2.7 are positive numbers in such an event.

There are two vertical lines, or decision axes, in the lone bottom panel of Figure 2.4. This is used in the two-tailed case, that is, when the direction of the difference between u_E and u_C is not specified in H_1. The entries in the last row are used as negative and positive numbers, respectively, when the left and right decision axes are used.

Several points may be said about the graphical representation of NHSTP. First, suppose that NHSTP is described in raw-score units (e.g., Column 1 of Table 2.5 or the right-most column of Table 2.6). As different sampling distributions are used in the one-sample and two-sample cases, the horizontal axis of the graphical representation of NHSTP should be given different labels for the two situations. Specifically, all the possible values of the

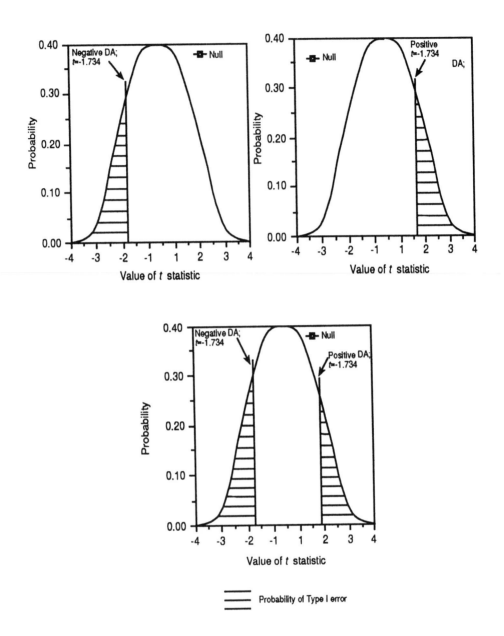

Figure 2.4   *The rationale of the one-tailed (top two panels) and the two-tailed (bottom panel) NHSTP*

Table 2.7   *The critical values of* t *for df = 18*

Level of significance (α) for one-tailed test	0.05	0.025	0.01	0.005
Level of significance (α) for two-tailed test	0.10	0.05	0.02	0.01
	1.734	2.101	2.552	2.878

sample mean are represented on the horizontal axis of the sampling distribution in the case of the one-sample test. The horizontal axis represents all the possible values of the difference between two means in the case of the two-sample test.

However, this distinction is not always observed in statistics textbooks. A common practice is to leave the horizontal axis of a diagram like Figure 2.1 or 2.2 without a label. However, the standard deviations of the two distributions are identified as standard error of the mean. Readers are left to assume that what is represented in the diagram are two sampling distributions of means, regardless of whether it is the one-sample or two-sample case. This misleading mode of representation is adopted for Figures 2.1 and 2.2 because they are commonly encountered representations (see, e.g., Darlington & Carlson, 1987, Figure 12.1, p. 314; Gravetter & Wallnau, 1966, Figure 8.10, p. 254). The misrepresentation is rectified in Figures 2.3 and 2.4, in which the horizontal axis represents all possible values of the *t* statistic (i.e., either sample mean or mean difference expressed in the appropriate standard error units).

Second, the rationale of NHSTP is adequately represented solely by the sampling distribution prescribed by H_0 (see, e.g., Hopkins, Hopkins & Glass, 1996, Figure 11.1, p. 192). To see the third point, it is necessary to recall that the graphical representation of the probability of Type II error (i.e., β) in Figure 2.1 requires the sampling distribution contingent on H_1, not on H_0. The inescapable conclusion is that Type II error plays no role in NHSTP, even if a representation like Figure 2.1 were correct (see Section 6.8).

2.7.2.   *Statistical Significance: Criterion and Meaning*

We learned from Section 2.5.3 above that the sampling distribution of differences is a theoretical probability density function which describes all possible values of the difference between two sample means, as well as their respective associated probabilities of occurrence *in the long run* by virtue of the *Central Limit Theorem*. The sampling distribution is used as a conceptual frame of reference for defining the criterion used to decide whether or not it is reasonable to explain the data in terms of chance influences [viz., $p(\text{Data}|H_0)$], a position found in the Fisherian perspective (see Row 4 of Table 2.3; see also Inman, 1994).

The bell shape of the sampling distribution of differences indicates that, if chance is responsible for the data, the majority of the possible values of the difference are around the difference between the means of the two parent populations (viz., zero in the present example; see Section 2.6.2). Although differences that differ from zero by a relatively extreme magnitude in either the positive or the negative direction are possible (call them 'extreme differences'), the probability of obtaining an extreme difference is progressively smaller the more extreme the difference.

The sampling distribution of differences is used, in its capacity as a probability density function, to provide an arbitrary, but well-defined, criterion for deciding whether or not to accept chance as the explanation of the data. This decision says nothing about the probability of the null hypothesis being true. The rationale of this choice begins with the recognition that it is impossible to exclude chance with absolute certainty as an explanation. Although it is unreasonable to treat frequent events in the same manner as infrequent events, it is reasonable to ignore an outcome if its probability of occurrence is small enough. The question is how small the probability of an infrequent event should be, *conditional upon* chance influences, before it becomes reasonable to reject chance as an explanation.

Some conventional values have been adopted for the purpose of excluding chance as the explanation of research outcome. They are the α values .05, .01 and .005 for most cases (see the column headings of Table 2.7). An α level (e.g., .05) is graphically represented by the area shaded with horizontal lines in Figure 2.4 (i.e. the probability of Type I error). The position of the decision axis is measured, in units of standard error, by its distance from the mean of the sampling distribution.

For the purpose of illustration, consider the top right panel of Figure 2.4. The critical value of t is 1.734 at the .05 level. It means that, if the decision axis is placed at 1.734 standard error units from the mean of the sampling distribution, the area to the right of the decision axis is 5% of the total area under the sampling distribution. The critical t value, 1.734, is the criterion for deciding whether or not to explain the data in terms of chance influences. This critical value is determined solely with reference to the null distribution. The criticism that NHSTP leads the researcher to emphasize Type I error at the expense of Type II error should be reconsidered in light of the fact that the 'alternative hypothesis' distribution plays no role at all in deciding the critical value. This reinforces the observation made in Section 2.7.1 that Type II error plays no role in NHSTP.

2.7.3. *A Decision or an Inferential Procedure?*

There is sometimes a dispute as to whether NHSTP is an inferential procedure or a statistical decision. The former is a Fisherian feature, whereas the latter is the Neyman–Pearson characteristic (see Row 3 in Table 2.3). As may be recalled from the upper right hand panel of Table 2.1 (see Section 2.2.2), the sufficient condition for H_0 is that the calculated t is smaller than 1.734. The complementary operationalization of 'not H_0' is that the calculated t is equal to, or larger than, 1.734 (viz., the upper left hand panel of Table 2.1). Hence, the major premises of the two conditional syllogisms are (a) '*If* Calculated $t \geq$ (criterion = 1.734), *then* not H_0' and (b) '*If* Calculated $t <$ (criterion = 1.734), *then* H_0' in the upper panel of Table 2.1. Of present interest is the binary decision made with reference to the criterion value (1.734 in this example). This supports the Neyman–Pearson view of a decision-making process. However, there is more to NHSTP than this binary decision.

Suppose that the difference between $\overline{X}_E$ and $\overline{X}_C$ gives rise to $t = 2.05$. A binary decision is made *vis-à-vis* the criterion (1.734 in this example) as to which one of the two conditional syllogisms in the upper panel of Table 2.1 to use. The decision is 'Calculated $t \geq$ (criterion = 1.734)', and this decision serves as the minor premiss for the conditional syllogism depicted in the top-left panel of Table 2.1. That is, NHSTP is an inferential process at this stage of the exercise.

The conclusion of the example is to exclude chance as the explanation of the data. The result is said to be statistically significant. That is, the meaning of statistical significance, or the rejection of H_0, is that the result is deemed unlikely if only chance influences are operating. It is deemed unlikely because the conditional probability, $p(\text{calculated } t|H_0)$ or $p(\text{Data}|\text{Chance})$, is smaller than the α level. Important to subsequent discussion of effect size (Chapter 5) and statistical power (Chapter 6) are the facts (a) that the H_0 distribution is the sampling distribution of differences whose mean difference is zero, and (b) that the H_1 distribution has no contribution to the concept of statistical significance. This state of affairs is contrary to Schmidt's (1994) assertion that the sampling distribution predicated on H_0 is irrelevant to the statistical decision.

In view of the relationship between the upper and lower panels of Table 2.1, it may be suggested that NHSTP consists of (a) a binary decision to choose between two alternative characterizations of the data (viz., 'Is the calculated t smaller than the criterion value?' – see Step 8 in Table 1.1). (b) an inference using the appropriate conditional syllogism (e.g., in the upper left panel of Table 2.1 if the calculated t is not smaller than the criterion value), and (c) a second inference utilizing a disjunctive syllogism (viz., the lower panel in Table 2.1). That is, the decision and inferential processes are carried out at successive stages of the exercise. In short, the decision-making procedure recognized in the Neyman–Pearson approach and the inferential procedure envisaged in the Fisherian approach (see Row 3 in Table 2.3) are not mutually exclusive because they refer to different steps in NHSTP. This state of affairs is contrary to the assertion that 'the hypothetico-deductive method . . . does not work with statistical inference' (Pollard, 1993, p. 456).[7]

2.8. Some Criticisms of NHSTP Reconsidered

It is now possible to reconsider a few criticisms of NHSTP relating to (a) the arbitrariness of the α level, (b) the incorrect interpretation of p, (c) the 'criterion-choice ambiguity', (d) the relationship between the α level and the

degree of evidential support offered by the data, and (e) the institutionalization of NHSTP.

2.8.1. The Arbitrariness of the α Level

Consider the upper-right panel of Figure 2.4. Locating the decision axis at
1.734 standard error units from the mean difference is an arbitrary decision.
Does this arbitrariness detract from the validity or the usefulness of the
concept of statistical significance? The issue of usefulness can easily be
settled. For example, the choice of the colours used in regulating traffic is
arbitrary. Yet, no one would doubt the usefulness and importance of this
arbitrary choice. The question of validity, on the other hand, is a more
difficult question. One way to deal with it is to consider it as a question about
justification. How can the arbitrariness of the criterion be justified?

One possible justification is that it is a well-defined criterion which is
meaningful and objective at the mathematical level. This is more satisfactory
than justifying the effect size convention with the tautological assertion,
'the proposed conventions will be found to be reasonable by reasonable
people' (J. Cohen, 1987, p. 13). Moreover, its meaning is independent of
the researcher's theoretical preference or commitment in the content area.
It is used as an unambiguous index of the strictness of the decision criterion
adopted by the researcher in rejecting chance as the explanation of the data.
It is not clear why '[significance] testing creates an illusion of objectivity' to
Schmidt (1994, p. 20).

A smaller α value indicates a stricter criterion. Hence, a .01-level decision
is a stricter criterion than a .05-level decision. However, this index of
criterion strictness says nothing about the truth or falsity of H_0. Nor is it
informative as to the degree of evidential support for the substantive
hypothesis offered by the data. Questions about the evidential support
offered by research data will be dealt with in Chapters 3 and 4. Meanwhile,
suffice it to say that the Neyman–Pearson characterization of the nature of
NHSTP, namely, the acceptability of H_0 in Row 2 of Table 2.3, seems a
more reasonable one than Fisher's characterization (i.e., to determine the
truth of a hypothesis).

2.8.2. The Meaning of the Associated Probability (p)

The probability associated with the test statistic (e.g., p of the calculated t) is
not available from the prepared statistical tables found in statistics test-
books. However, p is routinely provided by computer statistical packages. It
gives the exact placement of the calculated t value on the continuum of
theoretical t values. The present discussion shows why it is incorrect to treat
p as the probability that the results are due to chance.

The first reason is that 'associated probability' does not refer to the exact
probability of a particular test-statistic value. Instead, it refers to the
probability of a particular test-statistic value ($t = 2.05$ in our example) *plus*

the probabilities of all more extreme possible values (viz., $t = 2.56, 3.29$, etc.); that is, the 'tail area [of the sampling distribution] integral' (Oakes, 1986, p. 120). In other words p is a cumulative probability.

Second, at the same time, p is a conditional one; namely, it is a cumulative probability contingent on the acceptance of H_0. In other words, to treat p as the probability that the results are brought about by chance is to put the cart before the horse. This misinterpretation of the meaning of p is an example of misuse of NHSTP concepts identified by Gigerenzer (1993), as well as by many other critics. It may readily be avoided if H_0 is used properly as the antecedent of Proposition [P2-5].

2.8.3. The Criterion-choice Ambiguity

A calculated t of 2.05 is significant at the .05 level because it exceeds the critical value of 1.734 of a one-tailed test. However, had the α level been .01, the result would not be significant because the critical value of t at the .01 level is 2.552 (see Table 2.7). This criticism follows from the arbitrary choice of the α level, and it may be called the 'criterion-choice ambiguity' criticism. In view of this criticism, the Neyman–Pearson recommendation that the α value be set before data analysis is more satisfactory than the Fisherian practice (see Row 9 in Table 2.3). This means that the conditional probability value is the sole basis of quantifying the strictness of the researcher's decision made about the chance theory. The rationale of the decision rule is consequently not subject to the whim or some extra-statistical concerns of NHSTP users.

It is allowed in the Fisherian approach, on the other hand, to use the associated probability, p, to reject the null hypothesis. This state of affairs makes it possible for the researcher to reject the null hypothesis at $p = .048$ in one case, but at $p = .052$ in another. Some critics are sympathetic with this flexibility (e.g., J. Cohen, 1987; Oakes, 1986). However, this flexibility is gained by including some additional, yet undefined, criteria in making the decision. It is undesirable because the decision criterion is no longer a well-defined one. The more serious objection to this mode of making decisions is that H_0 is no longer a hypothesis contingent on chance influences alone because additional undefined criteria are included in making the decision. Consequently, it is no longer clear what rejecting or accepting H_0 means (see Section 5.3.3 below).

Is the 'criterion-choice ambiguity' criticism valid if the α level is determined before data analysis? The criticism is unanswerable at the logical level. However, it is a vacuous criticism for the following reason. Suppose that the calculated t is 2.72. The chance explanation can be rejected at the .01 level (one-tailed). However, the logical criticism may then be raised that the outcome is not significant at the .005 level at which the critical value is 2.878 (see Table 2.7). The important points are (a) that this criticism is raised after data analysis, and (b) that it is an open-ended criticism because it is possible to suggest the .001, .0005, .0001, etc., levels.

The vacuity of this logical critique may further be seen if the argument is turned around. Suppose that the calculated t is 1.698. It is not significant at the .05 level. In view of the 'criterion-choice ambiguity' criticism, it would be legitimate for the researcher to say that the result is significant at the .07 level. This is the case because the 'criterion-choice ambiguity' criticism suggests effectively a decision criterion of flexible strictness. However, the onus is on critics to justify the flexibility.

2.8.4. The Relationship between the α Level and Evidential Support

A more cogent rejoinder to the 'criterion-choice ambiguity' criticism may be offered if the distinction is made between the spirit of the criticism and the erroneous assumption about NHSTP which informs the criticism. One can accept the spirit without accepting the assumption of the criticism. Consider first the spirit of the criticism.

Recall that the critical $t_{(df=18)}$ value at the .05 and .01 levels are 1.734 and 2.552, respectively, for the one-tailed case. That is, the 'statistical significance' decision axis is placed at 1.734 and 2.552 standard error units, respectively, from the mean of the sampling distribution. Suppose that (a) $(\overline{X}_E - \overline{X}_C)$ is the same for Studies A and B, (b) the calculated t of Study A exceeds $t_{(df=18)} = 1.734$, but not $t_{(df=18)} = 2.552$, and (c) the calculated t of Study B exceeds $t_{(df=18)} = 2.552$. This state of affairs means that the standard error of Study B is smaller than that of Study A. This difference between Studies A and B at the statistical level is brought about by some non-statistical features of the studies, such as their designs, the tasks used in the two studies, the amount of training provided for the subjects, and the like.

In other words, the 'criterion-choice ambiguity' criticism is, in fact, raised to indicate some non-statistical doubts about the study. Hence, the spirit of the 'criterion-choice ambiguity' criticism is legitimate, namely, that Study A could have been conducted in a more satisfactory manner. More importantly, the criticism is a dissatisfaction with the strictness of the decision criterion, not with the fact that an arbitrary decision criterion is used.

Now consider the assumption of the 'criterion-choice ambiguity' criticism. The criticism is informed by the erroneous assumption about NHSTP that a result significant at the .01 level confers more support to a result significant at the .05 level. However, the following case may be made. To say that a decision criterion in Study A is less strict than that used in Study B is not to sa0 that the result of Study A is less valid than that of Study B.

The reason is that other important features of Study A (e.g., its design, stimulus materials, dependent variable, amount of training) are predicated on the assumption that the chosen level of strictness (viz., $\alpha = .05$) is acceptable before data collection. Had a stricter criterion been deemed necessary before data collection, the other features of the study would have been adjusted accordingly. For example, the subjects may be trained more extensively when a stricter criterion for the Type I error is called for. This is a decision at the level of data collection, not a statistical concern. In short, although the 'criterion-choice ambiguity' is a legitimate criticism of empirical research, it is not a criticism of NHSTP *per se* (see Sections 4.7.4 and 4.8). Of present interest is the debatable assumption about the relationship between evidential support and the level of significance.

It would make sense to say that the .01 level confers more evidential support than the .05 level only if it were true that (a) p is related to the probability of H_0 being true, and (b) H_0 (or its complement) is related to the substantive hypothesis. However, neither (a) nor (b) is true. That Point (a) is not true may be recalled from the discussion in Section 2.8.2 that the associated probability (p) is a probability conditional upon accepting H_0. Hence, p cannot inform the researchers as to the probability of H_0 being true. An answer to Point (b) is implicit in what has been said about H_0 in Section 2.6. It is shown that H_0 is the hypothesis about the effect of chance influences on the data-collection procedure. It says nothing about the substantive hypothesis or its logical complement, a point to be further developed in Section 3.2 (particularly Sections 3.2.4 and 3.2.5). Of interest here is that the difference between the .01 and .05 levels does not indicate differential degrees of support for the substantive hypothesis by the two respective sets of data. The difference is one about the strictness of the decision criterion used to reject the explanation in terms of chance influences.

2.8.5. The Institutionalization of NHSTP

A rough summary of NHSTP includes (a) a prescriptive conventional α level of .05, (b) a mechanical[8] decision rule of rejecting H_0 (viz., reject H_0 if the calculated test statistic is as extreme as, or more extreme than, the criterion), and (c) a rule of ignoring and discarding a non-significant result. Gigerenzer's (1993) view is that these features of NHSTP are routinely enforced to the extent that NHSTP becomes 'institutionalized,' and that a significant result is treated as a criterion of good research. (See also Danziger, 1990, for a similar view.) Consequently, nonsignificance appears a 'negative, worthless, and disappointing result' (Gigerenzer, 1993, p. 319). Moreover, this institutionalization breeds the erroneous view that NHSTP is objective. Suppose that the data yield statistically insignificant results. NHSTP users reject dogmatically the theory instead of questioning the data. Gigerenzer (1993) raises the logical question that the data may be the culprit.

As an indictment of the *misuse* of NHSTP, the spirit of Gigerenzer's (1993) argument is reasonable. However, it is not an argument against setting the α level at the conventional value before data analysis. Nor does it say why the mechanical binary choice between *chance* and *not chance* with reference to a well-defined criterion prescribed by the α level should be abandoned. These features (viz., being mechanical and being conventional) do render NHSTP objective, Gigerenzer's (1993) charge of dogmatism notwithstanding, for the following reason.

The crux of the matter is that there is a positive message in Gigerenzer's (1993) comment. It is a reminder that the conventional α level and the mechanical rule in NHSTP must be supplemented by the assurance that the data are properly collected. This is a consideration of the validity of the measurement made in the study and of the design of the study. In other words, Gigerenzer (1993) may be advising researchers to take into account design issues (viz., Campbell & Stanley's, 1966, inductive conclusion validity[9]), as well as issues relating to data analysis (i.e., Campbell & Stanley's, 1966, statistical conclusion validity) when assessing empirical research.

2.9.   Summary and Conclusions

Whether or not NHSTP is distinguishable from the scientific procedure is not a statistical issue. It will be dealt with in Chapters 3 and 4. A summary of the arguments developed in this chapter is presented with reference to the topics listed in Rows 2–11 of Table 2.3.

To begin with, NHSTP consists of (a) a binary decision, (b) an inferential procedure involving a conditional syllogism, and (c) a second inferential procedure involving a disjunctive syllogism. In other words, both the Neyman–Pearson 'decision' and Fisherian 'inferential' characterizations of the nature of NHSTP are true. However, they apply to different stages of NHSTP.

Specifically, the binary decision is whether or not the calculated test statistic is as extreme as, or more extreme than, the criterion. The outcome of the binary decision serves as the minor premiss of a conditional syllogism. The conclusion of the conditional syllogism serves as the minor premiss of the subsequent disjunctive syllogism in the event that H_0 is rejected. In other words, the Neyman–Pearson and Fisherian perspectives are not incompatible in terms of the objective and procedural nature of NHSTP.

The idea of inverse probability is that the researcher revises the probability of a theory being true in light of new data. Hence, random sampling becomes an exhortation, in the Neyman–Pearson perspective, to (a) carry out the data collection procedure repeatedly, and (b) revise the probability at the end of every trial. However, given (i) the theoretical nature of the sampling distribution of the test statistic, and (ii) that H_0 is the sufficient condition for using the sampling distribution of the test statistic with a specific parameter, the researcher is actually dealing with the conditional probability, $p(\text{Data}|H_0)$. That is, the Fisherian position in this regard is consistent with the mathematical foundation of NHSTP, but the Neyman–Pearson preference for the inverse probability is not. The sampling distribution of the test statistic gives NHSTP its probabilistic character in the sense that the associated probability (p) and the α level owe their meanings to the sampling distribution of the test statistic. This sampling distribution is about the data-collection condition, not the substantive hypothesis.

The null hypothesis is contingent on assuming that chance influences are responsible for the research outcome. The 'chance' assumption has not been properly acknowledged in most criticisms of NHSTP. Consequently, it is easy to lose sight of the facts that (a) the null hypothesis is an implication of the assumption about chance influences (i.e., [P2-4] in Section 2.6.2), and (b) the null hypothesis prescribes the choice of the sampling distribution of the test statistic (i.e., [P2-5] in Section 2.6.2). As will be seen in Chapter 3, H_1 is not the substantive hypothesis. Instead, it is the implication of the experimental hypothesis at the statistical level; and H_1 is the logical complement of H_0. Consequently, the Fisherian practice of treating H_0 (chance) and H_1 (qua an assertion of a departure from chance influences) as two mutually exclusive and exhaustive alternatives is correct.

The α level owes its meaning to the sampling distribution of the test statistic. This makes it possible to have a well-defined index for the researcher's strictness in rejecting or not rejecting chance influences as the explanation of data. To set the α level before data analysis is to exclude any other criterion in quantifying the researcher's strictness in rejecting (or not rejecting) H_0. In contrast, the Fisherian practice of using p to reflect the researcher's strictness in this matter is unsatisfactory because it introduces additional criteria in the decision making. This is undesirable because the additional criteria are not always explicit. Nor are they necessarily well-defined.

The rationale of NHSTP can be explained with reference to the sole sampling distribution contingent on H_0. As may be seen from Inman's (1994) historical account, Fisher was correct not to consider Type II error because it plays no role in NHSTP. The graphical representation of the probability of Type II error (i.e., β) requires the sampling distribution contingent on H_1. However, H_1 is implicated neither in NHSTP nor in the definition of Type II error. This state of affairs becomes important in the subsequent discussion of effect size in Chapter 5 and statistical power in Chapter 6.

At the level of statistics, to say that the result is statistically significant is to say that the calculated test statistic is as extreme as, or more extreme than, a criterion value. At the conceptual level, to report a statistically significant result is to say that the data have a low probability of occurring if chance influences are used to explain the data. This leads to the decision to exclude chance influences as the explanation. As to what the non-chance factor is, NHSTP is not helpful. It is necessary to turn to a different component of the research procedure. However, this limitation of NHSTP has not always been properly recognized as a non-statistical issue.

Notes

1. Represented on the horizontal axis of Figures 2.1 and 2.2 is the sample mean. While this representation is not incorrect in the case of the one-sample case NHSTP, it is an incorrect representation of the two-sample case NHSTP discussed in Section 1.2. The reason why it is

incorrect will become obvious in Section 2.7.1 below (see also Section 6.8). This incorrect representation is used in Figures 2.1 and 2.2 because it is the graphical representation commonly used to talk about the rationale of NHSTP, effect size or statistical power. It will be seen in Section 6.7 that 'Type II error' is being defined differently in Figure 2.1 and Table 2.2. The difference is not made explicit here so as to conform to a practice found among some critics of NHSTP.

2. An example that is even less justifiable than the replication fallacy is the following defence of the inter-personal expectancy effects. Commenting on the fact that 84 out of 242 studies were consistent with the expectancy effects hypothesis, Rosenthal (1973) said, 'But we must not reject the theory because "only" 84 studies support it; on the contrary. According to the rules of statistical significance, we could expect five percent of those 242 studies (about 12) to have come out as predicted just by chance' (p. 59). The replication fallacy is about literal replications of the same exercise. However, the 242 studies of the inter-personal expectancy effect are not literal replications of the same study. The statement in question is a misrepresentation of what statistical significance means.

3. There is, of course, the distinction between the related-sample and independent-sample cases. This distinction is ignored here for ease of exposition. This decision does not affect the logic of the illustration here. The independent-sample case is used in the present example.

4. The t statistic is not unlike a commonly used index of the effect size, namely, $d = (u_E - u_C)/\sigma_E$ (J. Cohen, 1987). This point will be raised again in Chapter 5.

5. The 'efficacious' characterization has different meanings in different types of experiment, a topic to be further developed in Sections 5.7.1 and 5.7.2 below.

6. The null hypothesis can be true even if it is used as a categorical proposition describing the substantive hypothesis. Specifically, it is said, 'Suppose an ESP experiment tests whether the thoughts of one person can influence the thoughts of another person in another room. The null hypothesis would be that there is no influence. If this null hypothesis cannot be correct, we can conclude that ESP exists before performing the experiment' (Frick, 1995, p. 132).

7. Pollard (1993) made this statement because he had chosen two debatable conditional propositions as the major and minor premises of the conditional syllogism (see Pollard, 1993, p. 456). Falk and Greenbaum (1995) also argue that it is inappropriate to characterize NHSTP as a deductive process. They have in mind the incorrect example used by Harshbarger (quoted by Falk & Greenbaum, 1995, p. 81).

8. The characterization, 'mechanical', was used by Gigerenzer (1993, p. 321). A rule or procedure is characterized as 'mechanical' when (a) it is well-defined, and (b) it guarantees the same result if it is followed properly, regardless of the rule-user or the circumstances in which the rule is applied. Rules in deductive logic are mechanical rules. It is important to note that 'mechanical' is not a derogatory characterization.

9. Cook and Campbell's (1979) 'internal validity' has two components, namely, statistical conclusion validity and internal validity. It is confusing to use the same name for a genus and a species of the genus. Cook and Campbell (1979) mean by the species *internal validity* the logical structure of the experiment. At the same time, the logical structure of the experiment is determined by the inductive rule underlying the experimental design (see Section 4.5). Consequently, Chow (1987a, 1992a) uses inductive conclusion validity to refer to Cook and Campbell's (1979) *internal validity*.

3

A Phenomenon and its Quartet
of Hypotheses

The logical relations among the various hypotheses implicated in the theory-corroboration experiment are made explicit so as to explain the meta-theoretical issues and distinctions implicit in some criticisms of NHSTP. The theory-corroboration, utilitarian, clinical and generality experiments are distinguished in terms of (a) the objective of conducting the experiment, (b) the nature and role of the substantive hypothesis and (c) the relationship between the substantive and experimental hypotheses. Issues about the putative superiority of the effect size over statistical significance are examined with reference to the four types of experiment. Also considered is the fact that statistical significance is not informative of the substantive impact of the research result.

3.1. Introduction

The example used in Section 1.2 is typical of the examples used in statistics textbooks to introduce the rationale of NHSTP. If its 'Method E' and 'Method C' are replaced by 'New Fertilizer' and 'Current Fertilizer', respectively, the example becomes one used in early agricultural research. Hence, the examples commonly used in statistics textbooks have been called 'agricultural model' experiments (Hogben, 1957; Meehl, 1978; Mook, 1983). The objective of the 'agricultural model' experiment is to discover or assess the practical importance of the manipulation in question (e.g., to use the new or the old fertilizer). Hence, such an experiment will be called the 'utilitarian experiment' in subsequent discussion because of its practical objectives (viz., discovery or assessment).

The general impression conveyed by the examples in statistics textbooks is that one finds in the utilitarian experiment all the essential features of experimentation. This is a wrong impression because something is amiss in the example. Specifically, Meehl's (1967) distinction between the substantive theory and the statistical hypothesis of an experiment is not obvious in the utilitarian experiment (see [P1-3] in Section 1.10).

The present discussion begins with the logical structure of the theory-corroboration experiment. Three more types of experiment are then distinguished. The differences so identified among the four types of experiment will be used to discuss the following issues: (a) the objective of experimentation, (b) the apparent interchangeability of the substantive and experimental hypotheses in the utilitarian experiment, (c) the contribution of the confidence-interval estimate and effect size in the theory-corroboration experiment, and (d) the substantive impact of experimental results. The terms 'theory' and 'hypothesis' both refer to a speculative account for a phenomenon. For this reason, the two terms will be used interchangeably in subsequent discussion even though 'theory' may have a more grandiose connotative meaning than 'hypothesis'.

3.2.   The Logical Relations among Various Hypotheses

The statistical alternative hypothesis, H_1, of NHSTP is often identified with the to-be-investigated substantive hypothesis in methodological or meta-theoretical discussions. This may be a remnant of Fisher's (1959, 1960) practice of identifying H_0 with the assertion, 'P does not exist', and H_1 with the assertion, 'P exists', where P refers to a phenomenon. This is misleading because diverse hypotheses at different levels of abstraction are implicated in the theory-corroboration experiment. What H_1 represents may be seen by considering the logical relationships among these various hypotheses with reference to Table 3.1.

3.2.1.   *Phenomenon and Theory*

Consider the fact that native speakers of English can understand and produce an infinite number of grammatical utterances, including those they have not heard before. This linguistic competence is the to-be-explained phenomenon. To Fodor (1975) and Miller (1962), a mental analogue of Chomsky's (1957) transformational grammar can explain the linguistic competence (see [P3.1.1] in Table 3.1). That is, linguistic competence (as the to-be-explained phenomenon) is temporally prior to the transformational grammar hypothesis at the psychological level. This temporal relationship may be represented as the 'phenomenon → hypothesis' sequence. As the phenomenon may be treated as observations or data prior to the hypothesis, the aforementioned sequence may also be called the 'prior data → hypothesis' sequence (Chow, 1992c). Furthermore, what is said in the hypothesis must be consistent with what is known about the phenomenon. That is, the prerequisite for the 'phenomenon → hypothesis' sequence is the phenomenon–hypothesis or prior data–hypothesis consistency, an important point in subsequent assessment of Bayesianism in Section 7.4.1.

Miller's (1962) explication of the transformational grammar at the psychological level is a substantive hypothesis in Meehl's (1978, 1990) terms. Moreover, this substantive hypothesis is a conjecture about a hypothetical mechanism in Popper's (1968b/1962) terms. How this substantive hypothesis

Table 3.1 *The logical relations among the to-be-explained phenomenon, theory, research hypothesis, experimental hypothesis and statistical hypotheses (alternative and null) in a theory-corroboration experiment*

Level of discourse To-be-explained phenomenon:	What is said at the level concerned The linguistic competence of native speakers of English	
Substantive hypotheses	The linguistic competence of native speakers of English is an analogue of the transformational grammar.	[P3.1.1]
Complement of *theory*	*The linguistic competence of a native speaker of English* *is not an analogue of the transformational grammar.*	*[P3.1.1']*
Research hypothesis	**If** [P3.1.1], **then** it is more difficult to process negative sentences than kernel sentences.	[P3.1.2]
Complement of *research hypothesis*	*If* −*[P3.1.1], then there is no difference in difficulty in* *processing negative and kernel sentences.*	*[P3.1.2']*
Experimental Hypothesis	**If** the consequent of [P3.1.2], **then** it is more difficult to remember extra words after a negative sentence than after a kernel sentence	[P3.1.3]
Complement of *experimental* *hypothesis*	*If not the consequent of [P3.1.2], then it is equally* *difficult to remember extra words after a negative and a* *kernel sentence.*	*[P3.1.3']*
Statistical alternative hypothesis	**If** the consequent of [P3.1.3], **then** H_1.*	[P3.1.4]
Statistical null *hypothesis*	*If not the consequent of [P3.1.2], then H_0.*†	*[P3.1.4']*
Sampling distribution of H_1	**If** H_1, **then** the probability associated with a difference between kernel and negative sentences as extreme as 1.729 standard error ($t_{df=19}$) units from an **unknown** mean difference is not known.	[P.3.1.5]
Sampling distribution of H_0	*If H_0, then the probability associated with a difference* *between kernel and negative sentences as extreme as 1.729* *standard error ($t_{df=19}$) units from a mean difference of* *zero is .05 in the long run.*	*[P3.1.5']*

*H_1 = mean of extra-sentence words recalled after negative sentences < mean of extra-sentence words recalled after kernel sentences.
†H_0 = mean of extra-sentence words recalled after negative sentences ≥ mean of extra-sentence words recalled after kernel sentences.

may be corroborated experimentally will be the topic of discussion in Chapter 4. Meanwhile, consider how an experimenter may come up with an experimental hypothesis used to test the substantive hypothesis of the psychological reality of the transformational grammar. Of interest is the fact that H_1 is neither the substantive nor the experimental hypothesis, a state of affairs not taken into account in criticisms of NHSTP.

3.2.2. Research Hypothesis – an Implication of the Theory

The hypothetical mechanisms implicated in the transformational grammar is not observable (e.g., the mental representation of the phrase structure and transformational rules). It is, hence, impossible to investigate the theoretical structures and functions underlying linguistic competence in the way observable objects or events are studied (e.g., a chair or a football match). To investigate these unobservable structures and functions, psychologists first deduce observable implications from the theory in a well-defined context. Consider the following example.

According to the transformational grammar (viz., the substantive hypothesis [P3.1.1] in Table 3.1), while a kernel sentence[1] is produced with the phrase-structure rules, a non-kernel sentence is the result of applying one or more optional transformational rules to a kernel sentence (Chomsky, 1957). For example, a negative sentence is the result of applying the negative transformation to a kernel sentence. An implication of the substantive hypothesis is that non-kernel sentences are more difficult to process than kernel sentences (e.g., in the form of imposing a heavier load on the working memory). In other words, an implication of the substantive hypothesis (e.g., the consequent of [P3.1.2] in Table 3.1) is a research hypothesis deduced from the substantive hypothesis (i.e., the antecedent of [P3.1.2]). This implication is theoretical in the sense that the research hypothesis says what it says by virtue of what is said in the substantive hypothesis. Moreover, the relation between the substantive and research hypotheses is implicative because the substantive hypothesis is not tenable if there is no evidential support for the research hypothesis.

3.2.3. Experimental Hypothesis – the Research Hypothesis in Context

The research hypothesis represented by the consequent of [P3.1.2] in Table 3.1 is not specific enough for conducting empirical research. Specifically, questions may be raised as to what the nature of the said processing activity is, or what the criterion of difficulty is. This problem of vagueness is solved by specifying (a) a well-defined experimental task in a specific situation, and (b) a specific dependent variable whose identity is independent of the substantive hypothesis. The result of this specification is the experimental hypothesis (viz., the consequent of [P3.1.3] in Table 3.1).

For example, Savin and Perchonock (1965) gave their subjects a sentence and eight additional words on every trial. The subjects' tasks were, first, to recall the sentence verbatim, and, second, to recall as many of the eight extra words as possible. The experimental manipulation was the number of grammatical transformations required for the generation of the surface structure of the sentence. While a kernel (K) sentence requires no transformation, a negative sentence (N), a passive sentence (P) or a question (Q) requires one transformation (the negative, passive or question transformation, as the case may be). A negative sentence in the passive voice (NP) requires two transformations (i.e., both the negative and passive transformations). A negative question in the passive voice (NPQ) requires three transformations from the kernel sentence (the negative, passive and question transformations). Savin and Perchonock (1965) used ten types of non-kernel sentence. However, for the present illustrative purpose, it suffices to concentrate only on kernel (zero transformation) and negative (one-transformation) sentences.

It follows from the transformational grammar that Savin and Perchonock's (1965) subjects had more things to remember for the verbatim recall of a negative sentence than a kernel sentence. Together with the auxiliary assumption that we have a short-term memory store whose capacity is limited (Miller, 1956), the antecedent of [P3.1.3] in Table 3.1 implies that, if more of this limited capacity is used to retain a sentence, less of the capacity is available for retaining the additional words in Savin and Perchonock's (1965) task. Hence, the antecedent of [P3.1.3] implies the experimental hypothesis represented by the consequent of [P3.1.3] in Table 3.1.

This experimental hypothesis is used to test the research hypothesis (viz., the consequent of [P3.1.2]). It is important to emphasize that the experimental hypothesis is not a simple re-phrasing of the research hypothesis, let alone of the substantive hypothesis. For example, the processing difficulty in question may be measured in a number of different ways. Specifically, it may be measured in terms of the number of errors made or the correct reaction time in the verbatim recall of the sentence. However, it is measured in terms of the number of extra words recalled in Savin and Perchonock's (1965) experiment. That is to say, the same research hypothesis may give rise to diverse experimental hypotheses in different experimental contexts. This state of affairs will become important when meta-analysis is assessed in Section 5.8.3 below.

3.2.4. The Statistical Alternative Hypothesis (H_1)

Important to the present discussion is the fact that an experimental hypothesis like the consequent of [P3.1.3] in Table 3.1 is not amenable to statistical analysis. Consequently, the implication of the experimental hypothesis is required at the statistical level. It is this implication, not the experimental hypothesis itself, that is the statistical alternative hypothesis (H_1) (i.e., the consequent of [P3.1.4] in Table 3.1). H_1 in this experiment reads:

> The mean number of additional words recalled after negative sentences is smaller than the mean number of additional words recalled after kernel sentences [i.e., H_1: $(u_{negative} - u_{kernel}) < 0$].

It is thus shown that the statistical implication of the experimental hypothesis is phrased in terms of the independent and dependent variables of the experiment. Furthermore, this implication of the experimental hypothesis is a prescription for this experiment in the sense that it stipulates what the evidential data should[2] be like at the statistical level in order for the experimental hypothesis to be tenable at the conceptual level.

The experimental hypothesis encapsulates the experimental task, the test conditions and the data-collection procedure. It is temporally prior to the availability of the experimental data. Hence, in contrast with the 'prior data → hypothesis' sequence noted in Section 3.2.1, this relationship between the experimental hypothesis and the experimental data may be characterized as the 'hypothesis → evidential data' sequence (Chow, 1992c).

It might be recalled from Section 2.5 that the mathematical basis of NHSTP is the sampling distribution of an appropriate test statistic. In the case of the present example, the tool required is the sampling distribution of differences. It is uniquely defined by the mean difference, the standard error of the difference and the degrees of freedom. As may be seen from the consequent of [P3.1.5] in Table 3.1, H_1 is not useful because it does not specify the mean difference when kernel and negative sentences are compared.[3] Consequently, it is not possible to determine the to-be-used sampling distribution of differences with reference to H_1. This difficulty leads to the statistical null hypothesis (H_0).

3.2.5. The Statistical Null Hypothesis (H_0)

The logical complement of the present H_1 is that the mean number of extra-sentence words recalled after negative sentences is equal to, or larger than, the mean of extra-sentence words recalled after kernel sentences. As has been noted in Section 2.2.5, whatever can be rejected on the basis of a zero difference (viz., $u_{negative} = u_{kernel}$) in the present situation can also be rejected on the basis of a positive difference (viz., $u_{negative} > u_{kernel}$). That is, for the purpose of making the statistical decision, the directional part of H_0 (i.e., the *larger than zero* component in our example) can be justifiably ignored because it is effectively subsumed under the non-directional part of H_0 (i.e., the *is equal to zero* component in our example).

Consider the utilities of H_0. First, the decision to accept H_1 may now be made via a decision made about H_0 by virtue of the disjunctive syllogism described in Section 2.2.3. Second, identifying H_0 as a statement of no difference makes it possible to specify the sampling distribution of differences whose mean difference is zero. It is for this reason that Proposition [*P3.1.5'*] in Table 3.1, whose antecedent is H_0, is used in tests of significance instead of Proposition [P3.1.5].

The introduction of H_0 in the present argument may seem contrived; it is introduced because H_1 is not useful. After all, H_0 has been said to be an imperfect theoretical statement that serves as a straw man (Bakan, 1966; Mosteller & Bush, 1954). That this is an incorrect view of H_0 may be seen from the entries in italics in Table 3.1 (viz., Propositions [*P3.1.1'*], [*P3.1.2'*], [*P3.1.3'*], and [*P3.1.4'*]). Beginning with the complement of the substantive hypothesis as the first premise ([*P3.1.1'*]), the consequent of

[P3.1.3'] follows the same sequence of derivations from *[P3.1.1']* as does the consequent of [P3.1.3] from [P3.1.1]. Hence, *[P3.1.3']* is not a contrived proposition if *[P3.1.1']* is not an arbitrary assertion. Proposition *[P3.1.1']* is not arbitrary by virtue of the fact that it is a logical complement of the substantive theory [P3.1.1]. This route of deduction is a good rejoinder to Fisher's (1959; see Inman, 1994) or Rozeboom's (1960) assertion that the choice of the null hypothesis is arbitrary.[4]

The route of deduction from *[P3.1.1']* through *[P3.1.4']* suggests that it is oversimplifying, and hence misleading, to say that H_0 (as a categorical proposition[5]) is merely the hypothesis of no difference. A proper interpretation of H_0 should take into account that (a) it is a statement made contingent on the assumption that the only source of variation in the data is chance influences, (b) it is a statement of no difference between two well-defined populations (viz., defined in terms of the data-collection procedure), and (c) the two populations involved are identical in all aspects but one (see Section 2.6.2).[6] This interpretation is made necessary by the facts that (a) *[P3.1.3']* is the logical complement of [P3.1.3], (b) the consequent of [P3.1.3] is the experimental hypothesis, and (c) the specific nature of the experimental condition is taken into account in formulating the experimental hypothesis, as well as its experimental prescription.

3.2.6.   Research Refinement – Statistical or Non-statistical

It has been shown that the experimental hypothesis is derived from the substantive hypothesis by a two-step implicative process (i.e., from [P3.1.1] to [P3.1.3] via [P3.1.2] in Table 3.1) in a particular experimental context. The said context includes the nature of the experimental task, the choice of independent and dependent variables, stimulus materials and the mode of stimulus or task presentation, and the like. Hence, the implicative relationship between the substantive and experimental hypotheses may be improved by determining judiciously these features of the experiment. This is how one may understand Meehl's (1967) idea of improving the implicative relationship between the substantive and experimental hypotheses. Of interest to the present discussion is the fact that these considerations are non-statistical in nature. The more interesting consequence of improving the data-collection procedure of the experiment is to ensure that the null hypothesis is correct (see Section 2.6.2, and Section 3.5 below).

3.3.   Chance versus Non-chance Factors

It is necessary to make explicit the meaning of the consequent of *[P3.1.3']* in Table 3.1. First, suppose that the grammar of the sentence has no effect on performance. It means that the performance on negative and kernel sentences represents two respective populations whose means are identical. Second, any observable difference in performance between negative and

Table 3.2   *The statistical null hypothesis (H_0) and the statistical alternative hypotheses (H_1) as components of conditional propositions*

Where in Table 3.1	Conditional proposition		Refer to
[P3.1.4']	**If** chance, **then** H_0.	[P3.2.1]	[P2-4] in Section 2.6.2
[P3.1.4]	**If** not chance, **then** H_1.	[P3.2.2]	
	If H_0, **then** the test statistic is distributed as a *sampling distribution of the difference* whose mean difference is zero.	[P3.2.3]	[P2-5] in Section 2.6.2

kernel sentences on any particular occasion is the result of chance influences (see Section 2.6.2).

This appeal to chance influences justifies the use of the sampling distribution of differences with a mean difference of zero (see [P3.2.3] in Table 3.2). Hence, *[P3.1.4']* in Table 3.1 is represented by [P3.2.1] in Table 3.2, in which the antecedent of the conditional proposition is specified explicitly as 'chance'. This practice can be traced back to Neyman and Pearson (1928). Two consequences follow from this assumption: (a) the experimental subjects' performance on the two kinds of sentence on a particular occasion is not necessarily identical, and (b) the consequent of Proposition [P3.1.3] in Table 3.1 should be interpreted as saying that an observed difference in performance between negative and kernel sentences is the result of some non-chance factors. Hence, Proposition [P3.1.4] in Table 3.1 is rephrased as [P3.2.2] in Table 3.2. Propositions [P3.2.1] and [P3.2.3] in Table 3.2 are foreshadowed by [P2-4] and [P2-5], respectively, in Section 2.6.2.

The emphasis on the distinction between chance and not chance in the respective antecedents of [P3.2.1] and [P3.2.2] in Table 3.2 is important. It iterates that the statistical null hypothesis is statistical (Falk & Greenbaum, 1995) because it describes what events are like if they are random, or chance, occurrences (Hogben, 1957). It also brings up the important differences between (a) competing alternative explanatory hypotheses at the conceptual level and (b) the multiple alternative statistical hypotheses at the numerical level. Proposition [P3.2.2] in Table 3.2 also makes it clear that H_1 is not synonymous with non-chance influences. H_1 is the consequence of a particular exemplification of non-chance influences.

3.4.   Alternative Hypotheses – Statistical versus Conceptual

Some criticisms of NHSTP arise because both its fans and foes use the term 'alternative hypothesis' to refer to two different things, namely, a statistical alternative hypothesis and an alternative explanatory hypothesis. Although these hypotheses belong to different levels of abstraction, they are treated as though they are interchangeable in the debate about NHSTP.

3.4.1. Statistical Alternative Hypothesis (H_1)

Neyman and Pearson (1928) made the assertion that, in addition to asking questions about Population P with parameter μ, it was also necessary to ask questions about Populations P' with parameter μ', P'' with parameter μ'', and the like (call it the Multiple-H_1 Assumption). The essence of this assumption is that there are, in fact, multiple numerical alternatives to H_0 (see also Rozeboom, 1960). That is, in addition to the sampling distribution based on H_0, critics envision numerous sampling distributions with a non-zero mean difference based on numerous H_1's.

However, this assumption should have no bearing on NHSTP because it requires only the sampling distribution of differences with a mean difference of zero (see Section 2.7.1). Why is the emphasis on alternative sampling distributions of differences with a non-zero mean difference? The answer may be the fact that the term 'alternative hypothesis' is also used in another sense, albeit at a different level of discourse.

3.4.2. Conceptual Alternative Hypothesis

Given any to-be-explained phenomenon, there are alternative explanatory theories at the conceptual level (Popper, 1968a/1959, 1968b/1962). In actual fact, different psychologists often explain the same phenomenon with various substantive hypotheses. Moreover, diverse hypothetical structures or functions are postulated in various theories. For example, some psychologists prefer Fillmore's (1968) case grammar to Chomsky's (1957) transformational grammar. Other psychologists may find Yngve's (1960) 'Depth' model a more satisfactory explanation of linguistic competence. Consequently, these various substantive hypotheses lead to different research and experimental hypotheses (similar to the schema depicted in Table 3.1). As these experimental hypotheses may implicate different independent and dependent variables in diverse task situations, they may lead to qualitatively different H_1's.

It may be seen thus that at the conceptual level, given any phenomenon, multiple explanatory accounts may be offered (call this the Reality of Multiple Explanations). However, Neyman and Pearson (1928), as well as many other critics, did not make the distinction between the Multiple-H_1 Assumption and the Reality of Multiple Explanations. That the distinction should be made may be illustrated with reference to Table 3.3.

3.4.3. Two Types of Alternative Hypothesis Compared

Consider first the 'Multiple-H_1 assumption' column in Table 3.3. In terms of the number of extra words recalled, the H_1 of the modified version of Savin and Perchonock's (1965) experiment is in Row [a]. In view of the

Table 3.3 *The distinction between* statistical alternative
hypothesis *and* alternative explanatory hypothesis

Multiple-H_1 assumption	Reality of multiple explanations
[a] H_1: $(u_{negative} - u_{kernel}) < 0$	[i] H_1: $(u_{negative} - u_{kernel}) < 0$
H_0: $(u_{negative} - u_{kernel}) \geq 0$	H_0: $(u_{negative} - u_{kernel}) \geq 0$
[b] H_1': $(u_{negative} - u_{kernel}) = -3$	[ii] H_1': $(u_{ca} - u_a) > 0$
H_0': $(u_{negative} - u_{kernel}) \geq 0$	H_0': $(u_{ca} - u_a) \leq 0$
[c] H_1'': $(u_{negative} - u_{kernel}) = 5$	[iii] H_1'': $(u_{n-p} - u_{n-n}) \neq 0$
H_0'': $(u_{negative} - u_{kernel}) = 0$	H_0'': $(u_{n-p} - u_{n-n}) = 0$

ca = counter-agent; a = agent.
n–p = negative sentence with positive meaning.
n–n = negative sentence with negative meaning.

Multiple-H_1 assumption, two additional alternatives to H_0 are shown in
Rows [b] and [c] in that column. Each one of these statistical alternative
hypotheses is a point-prediction. They are numerical alternatives. However,
for the Multiple-H_1 assumption to be synonymous with the Reality of
multiple explanations, it is necessary to show something like the following:
Alternative [a] is an implication of the transformational grammar, Alternative [b] is derived from the case grammar, and Alternative [c] follows
logically from Yngve's (1960) 'Depth' model. Ironically, the researcher
should be very unhappy about the three theoretical alternatives if this were
the case, for the following reason.

Such a state of affairs occurs when the three numerical alternatives are
alternative outcomes in the same experimental context (e.g., the same
independent variable is manipulated, as indicated by the subscripts used in
the Multiple-H_1 assumption column). This state of affairs occurs when there
is no qualitative difference among the theoretical structures or mechanisms
envisaged in the three hypotheses. This means that the three hypotheses
give qualitatively the same prescription in a well-defined task context.
Consequently, they do not differ in terms of how well they explain our
linguistic competence at the conceptual level. In other words, the three
hypotheses are merely variations of the same genre under such situations.
The choice among the three alternatives becomes a non-theoretical one. In
what sense does the quantitative difference in question matter if it does not
make any difference at the explanatory level?

Consider now the 'Reality of multiple explanation' scenario. To begin
with, additional independent variables generally have to be added to the
experiment in order to test the multiple explanatory hypotheses simultaneously (viz., the case grammar and the 'Depth' model, in addition to the
transformational grammar). For example, it becomes necessary to manipulate Type of Case or Sentence Modality (e.g., agent versus counter-agent) in
order to test the case grammar. The depth of the sentence structure would
have to be manipulated if the 'Depth' model were being tested. For
example, it may be necessary to manipulate the type of negative sentences

being used (a negative sentence with a positive meaning, say, versus one with a negative meaning).

A prerequisite of a successful theory-corroboration experiment is that different experimental prescriptions are implied by the qualitatively different theories. For example, the Reality of multiple explanations is validly exemplified if something like what is depicted in the 'Reality of multiple explanations' column in Table 3.3 is observed. Specifically, the prescription of the transformational grammar is in Row [i]. The case grammar prescribes H_1' in Row [ii]. The 'Depth' model prescribes H_1'' in Row [iii]. As may be seen from the subscripts of the various means, the three different statistical alternative hypotheses are the implications of their respective experimental hypotheses at the statistical level.

3.4.4. A Triad of Hypotheses: the Conceptual Hypothesis, H_1 and H_0

Two things need be emphasized. First, multiple conceptual alternative hypotheses give rise to their respective statistical alternative hypotheses. Second, the multiple statistical hypotheses (e.g., H_1, H_1' and H_1'') are not alternatives to one single H_0. They have their own respective null hypotheses (H_0, H_0' and H_0'', respectively) even though these null hypotheses are all numerically equal to zero. They are equal to zero under different conditions (viz., different independent variables are used to test the three conceptual alternative hypotheses). This is indicated by the fact that three different conceptual hypotheses implicate different independent variables (sentence-type, case, and negative-type, respectively, as may be seen from the subscripts in Table 3.3). In other words, each of these multiple null hypotheses describes what chance variations are like under its own unique set of conditions (negative versus positive sentences in the case of the transformational grammar, counter-agent versus agent sentences for the case grammar, and negative sentences with positive meaning versus negative sentences with negative meaning for the 'Depth' model).

While this situation satisfies the Reality of multiple explanations, it contradicts the Multiple-H_1 assumption. It can now be seen why the sort of exact point-prediction suggested by Meehl (1967) is not relevant to choosing among contending theories (see endnote 3). The choice among qualitatively different conceptual theories is made in terms of some qualitative criteria which are satisfied when NHSTP is used.

In short, the differences among the experimental expectations prescribed by diverse alternative substantive hypotheses at the conceptual level are *not* a matter of numerical differences such as $u_1 = 5$, $u_2 = 10$, $u_3 = 15$ and the like. Consequently, the exclusion of unwarranted alternative hypotheses at the conceptual level also is *not* a choice among numerically different H_1's (Chow, 1989), but rather a conceptual endeavour for which conceptual and methodological rigour is important (Chow, 1991a, 1991b). The difference between the conceptual alternative hypothesis and the statistical alternative hypothesis (H_1) renders it inappropriate to identify H_1 with an experimental (let alone a research or a substantive) hypothesis.

3.5. 'H_0 is never true' Revisited

To many critics, one of the fatal flaws of NHSTP is that it is based on the null hypothesis, which is never true (see Section 1.3). Serlin and Lapsley's (1985, 1993) 'Good-enough Principle', as well as Frick's (1995) 'zero-as-special-number' thesis, may be seen as attempts to mitigate this threat to the validity of NHSTP. As may be recalled, this putative flaw of NHSTP is also responsible for the methodological paradox identified by Meehl (1967) (see Section 1.11). However, it has been shown in Section 2.6.2 that the 'H_0 is never true' criticism of NHSTP is debatable.

What should be emphasized is that H_0 is not a description of the to-be-studied phenomenon. Neither is H_0 the substantive hypothesis or its complement. As may be recalled from Section 2.6.2, the null hypothesis, $u_E - u_C = 0$, is postulated in the context of (a) two data-collection conditions that are identical in all aspects but one (a point to be developed further in Section 4.5), and (b) the assumption that the aspect in which the two conditions differ is immaterial to the parameter of interest (viz., $u_E - u_C$).

In other words, it is not inconceivable that H_0, as a categorical proposition, is true because it may be possible to control properly the data-collection procedure in the experimental approach. As a matter of fact, a reason why it is necessary to plan the experiment before carrying it out is to ensure that the data-collection procedure is compatible with what is asserted in the null hypothesis. In short, the assertion 'H_0 is never true' is by no means as self-evident as is assumed by some critics of NHSTP.

There is a second reason why the 'H_0 is never true' criticism of NHSTP is debatable. As may be recalled from *[P3.1.4']* and *[P3.1.5']* in Table 3.1, H_0 is not used as a categorical proposition in NHSTP. It is used twice as a component of a conditional proposition (a point hinted at in Meehl, 1967; a point also made in Section 2.6.2; see Propositions [P2-4] and [P2-5]). This may be seen more readily from Table 3.2. H_0 serves as the condition necessary for accepting chance variations as the explanation in [P3.2.1]. Moreover, it serves as the sufficient condition in which a particular sampling distribution is used in [P3.2.3].

Seen in this light, the issue about the null hypothesis is not whether or not H_0 is true: the issue is whether or not it is justified to entertain the conditional propositions, [P3.2.1] and [P3.2.3] in Table 3.2. As will be shown in Section 4.5, the presence of the appropriate kinds of control in the experiment provides the justification for using H_0 as the consequent of the conditional proposition [P3.2.1]. Using H_0 as the antecedent of Proposition [P3.2.3] is justified by the Central Limit Theorem at the mathematical level. This is the reason why the 'H_0 is never true' criticism of NHSTP is misguided. An implication of this realization is that the methodological paradox identified by Meehl (1967) in Section 1.11 is not as persuasive as it first seems. Moreover, Serlin and Lapsley's (1985, 1993) 'Good-enough Principle' seems unnecessary.

3.6. Three Additional Types of Experiment

It has been shown in Section 3.2 that, unlike in the case of the utilitarian experiment (see Section 1.2), the distinction between the substantive and statistical hypotheses is readily seen from the logical relationships among the various hypotheses found in the theory-corroboration experiment (see Table 3.1). Does this mean that the utilitarian experiment has a different logical structure? At the same time, there is a third type of experiment which is conducted to identify or classify phenomena (called the 'clinical' experiment in subsequent discussion). There is also a fourth type of experiment whose objective is to establish the generality of the functional relations between variables (called the 'generality' experiment in subsequent discussion). What is the logical structure of the clinical or generality experiment like? As the distinction among these four types of experiment (the theory-corroboration, utilitarian, clinical and generality experiments) is relevant to criticisms of NHSTP, it is necessary to describe the logical structure of these various types of experiment.

3.6.1. The Utilitarian Experiment

Consider the following utilitarian experiment. Suppose that the staff of a psychology department are not happy with their students' understanding of statistics. They wish to assess the effectiveness of a new method of teaching statistics, Method E. One random group of students is taught with Method E and another with the old method, Method C. The question is whether or not Method E is more efficacious than Method C.

The logical relations implicated in the utilitarian experiment is shown in Table 3.4. That the distinction between the substantive and the statistical hypotheses is not explicit may be seen readily by comparing Proposition [P3.4.1]–Proposition [P3.4.4] in Table 3.4 to Proposition [P3.1.1]–Proposition [P3.1.4] in Table 3.1.

As may be seen from Table 3.4, the substantive hypothesis is not an explanatory theory to answer a 'Why' question. Instead, it is a pragmatic or utilitarian hypothesis to deal with an 'Is there an effect?' question or a question about 'uncertain quantity' (see Phillips, 1973). As no hypothetical mechanism is implicated in [P3.4.1], the consequent of [P3.4.2] is a simple paraphrase or rephrasing of [P3.4.1]. That is, the research hypothesis is synonymous with the substantive hypothesis. At the same time, the pragmatic nature of the research question stipulates what the experimental manipulation should be. Hence, the levels of the to-be-manipulated variable have to

Table 3.4 *The logical relations among the to-be-investigated phenomenon,
pragmatic, research and experimental hypotheses of the utilitarian
experiment*

Level of discourse	What is said at the level concerned	
To-be-investigated phenomenon:	A dissatisfaction with students' current understanding of statistics	
Substantive (pragmatic) hypothesis	Method E is more effective than Method C.	[P3.4.1]
Research hypothesis	*If* [P3.4.1], *then* Method E produces better understanding than Method C.	[P3.4.2]
Experimental hypothesis	*If* the consequent of [P3.4.2], *then* students taught with Method E have higher scores than those taught with Method C.	[P3.4.3]
Statistical alternative hypothesis	*If* consequent of [P3.4.3], *then* H_1.*	[P3.4.4]
Sampling distribution of H_1	*If* H_1, *then* the probability associated with a difference between Methods E and C as extreme as 1.729 standard error units from an **unknown** mean difference is not known (assuming df = 19).	[P3.4.5]
Sampling distribution of H_0	*If* H_0,† *then* the probability associated with a difference between Methods E and C as extreme as 1.729 standard error units from a mean difference of zero is .05 in the long run (assuming df = 19).	[P3.4.5']

*H_1 = mean of Method E > mean of Method C.
†H_0 = mean of Method E ≤ mean of Method C.

be Method E and Method C, the very things mentioned in the substantive
hypothesis. This state of affairs may be responsible for the following
assertion:

> Indeed, the crucial distinction between a statistical hypothesis and a substantive
> theory often breaks down. To perform a significance test a substantive theory is
> not needed at all; the result . . . is that many 'substantive theories' are statistical
> hypothesis (with high *a priori* plausibility). (Oakes, 1986, p. 42; the emphasis in
> the original)

[Quote 3-1]

In other words, the experimental hypothesis is the operationalization of the
research hypothesis, and thus of the substantive hypothesis. As this operationalization is amenable to statistical analysis, the statistical alternative
hypothesis is no different from the experimental hypothesis. Consequently,
Meehl's (1967) distinction between substantive hypothesis and statistical
hypothesis is not obvious in the case of the utilitarian experiment.

It may seem that the effect size is more important than statistical
significance or the associated probability, *p*, in the case of the utilitarian

experiment. Suppose that Method E is significantly more efficacious than Method C. At the same time, Method E is more expensive to use than Method C. At the pragmatic level, it may be necessary to do a cost–benefit analysis on the basis of the magnitude of J. Cohen's (1987) index of effect size, $(u_E - u_C)/\sigma_E$. The question arises as to whether or not the benefit of using Method E is 'large enough' to justify the additional expenses. Knowing the effect size is important under such circumstances.

None the less, what needs to be said is that the cost–benefit issue is not a statistical problem. By itself, even the effect size is not helpful. There must be a non-statistical criterion for utilizing the effect size in the cost–benefit analysis. Be that as it may, an interesting question arises. What should the staff of the psychology department do if the result is not statistically significant, and yet the effect size, $(u_E - u_C)/\sigma_E$, is deemed 'large enough' in terms of the non-statistical criterion to justify adopting Method E?

It may not be advisable for the psychology department to adopt Method E for the following reason. To say that the result is statistically insignificant is to say that chance influences cannot be ruled out as an explanation of the putative efficacy of Method E, regardless of the magnitude of the effect. It means that the data-collection procedure is wanting, as indicated by the size of standard error of the difference. The solution is to collect new data (*not* additional data) with an improved data-collection procedure. As will be seen in Chapters 4 and 6, what is suggested here goes beyond choosing the appropriate sample size.

However, critics seem to find it unnecessary to collect new data to resolve the ambiguity. Under such circumstances, the issue becomes how to resolve the conflict between the statistical criterion underlying NHSTP and the non-statistical criterion underlying the cost–benefit analysis. If the psychology department is prepared to weigh the non-statistical criterion more heavily than the statistical decision, it may be asked why statistics is used at all in the first place.

3.6.2. The Clinical Experiment

The objective of the clinical experiment is a hybrid of pragmatic and explanatory concerns. Unlike the utilitarian experiment, the clinical experiment is carried out to determine if a particular phenomenon has occurred or if the phenomenon belongs to a particular theoretical category. For example, it may be asked whether or not a group of individuals with Characteristic C_1 belong to Category K (e.g., being extroverts). It is assumed that membership of Category K is indicated by the behavioural criteria C_1, C_2, ..., C_n. As may be seen from Table 3.5, a theory is implicated if Category K is theoretically defined (i.e., [P3.5.1] in Table 3.5). Moreover, a theoretical implication is deduced from the substantive hypothesis to be used as a research hypothesis (see the consequent of [P3.5.2] in Table 3.5).

However, unlike the theory-corroboration experiment, the substantive

Table 3.5　*The logical relations among the clinical question, substantive, research, and experimental hypotheses of the clinical experiment*

Level of discourse	What is said at the level concerned	
The clinical question:	A group of subjects with Characteristic C_1: do these individuals belong to Category K?	
Substantive hypothesis	Individuals belonging to Category K show Characteristics $C_2, \ldots, C_n$ in addition to Characteristic C_1.	[P3.5.1]
Research hypothesis	*If* [P3.5.1], *then* the subjects have Characteristic C_2.	[P3.5.2]
Experimental hypothesis	*If* the consequent of [P3.5.2], *then* the mean of the subjects on the C_2 task is larger than the mean of subjects who do not belong to Category K.	[P3.5.3]
Statistical alternative hypotheses	*If* the consequent of [P3.5.3], *then* **H_1**.*	[P3.5.4]
Sampling distribution of H_1	*If* H_1, *then* the probability associated with a difference between K and non-K subjects as extreme as 1.729 standard error units from an **unknown** mean difference is not known (assuming df = 19).	[P3.6.5]
Sampling distribution of H_0	*If* H_0,† *then* the probability associated with a difference between K and non-K subjects as extreme as 1.729 standard error units from a mean difference of zero is .05 in the long run (assuming df = 19).	[P3.5.5′]

*H_1 = mean of K subjects > mean of non-K subjects.
†H_0 = mean of K subjects ≤ mean of non-K subjects.

theory is assumed to be true in the clinical experiment. The substantive theory itself is not threatened even if the experimental outcome is inconsistent with the consequent of [P3.5.2] in Table 3.5. In such an event, the conclusion in the present example is simply that the group of individuals are not extroverts. The assumption is made that the phenomenon (i.e., the behaviour of the group of individuals in question) is explained if its category membership is identified.

At the same time, the experiment may also serve a pragmatic function, namely, to obtain the information necessary for deciding how to deal with the groups of individuals in question (e.g., to use one method of training rather than another). This utilitarian function is aided by the experimental hypothesis (i.e., the consequent of [P3.5.3] in Table 3.5) which is an operationalization of Characteristic C_2. The nature of the 'C_2 task' mentioned in [P3.5.1] in Table 3.5 may be very different from Characteristic C_1 or C_2. Like the theory-corroboration experiment, the substantive and statistical hypotheses are not the same.

Suppose that Characteristic C_2 is assessed by evaluating $(u_E - u_C)/\sigma_E$ where E and C are the to-be-assessed and norm groups, respectively, of the clinical experiment. The group is categorized in one way if $(u_E - u_C)/\sigma_E$ is positive, and in another way if $(u_E - u_C)/\sigma_E$ is not positive. Statistical significance is all that is required under such circumstances. The effect size may become relevant only if, after the diagnosis, the remedy called for is determined by the exact magnitude of the effect. Once again, the issue becomes a non-statistical one. The reservations about the role of the effect size in the utilitarian experiment expressed in Section 3.6.1 above are also applicable here.

3.6.3. The Generality Experiment

The term 'theory' or 'hypothesis' is used in Section 3.2 to refer to a speculative account which, if true, can explain a phenomenon. However, 'theory' or 'hypothesis' is sometimes used to refer to an empirical statement describing the functional relationship between observable variables (e.g., Earman, 1992; MacKay, 1993). Experiments are conducted to test the generality of various functional relationships. As may be seen from Table 3.6, it is necessary to describe the logical structure of this type of experiment in a slightly different way. However, this slight departure does not affect the rationale of experimentation to be described in Chapter 4.

There is nothing to explain in this type of research. Instead, it may be more fruitful to talk about its instigating observations. For example, it is observed that students who frequently rehearse newly learned words can remember the words better; choirs that practise more frequently sing better than those that rarely practise; tennis players who practise diligently win more matches than those who seldom practise. Having observed many instances of these events, the researcher abstracts a common features among these events (see Proposition [P3.6.1] in Table 3.6).

This abstraction is descriptive of events seen so far (viz., Skinner's, 1938, empirical laws). The researcher may speculate that all events involving the same two variables (practice and performance) would show the same regularity. This leap of faith is a generalization (viz., Hull's, 1943, empirical generalization). The result of such an act of induction by enumeration is represented by Proposition [P3.6.2] in Table 3.6. It is the research hypothesis of the generality experiment because the objective of conducting the experiment is to test whether or not [P3.6.2] is true.

Suppose Task T_1 is one that has not been observed before. It is a candidate for testing the generality of [P3.6.2]. Although the task used is new, the two variables used (i.e., practice and performance) are the same as those implicated in abstraction [P3.6.1] and generalization [P3.6.2]. Consequently, the experimental hypothesis is indistinguishable from the research hypothesis.

The generality of the empirical function refers to the range of experimental situations to which the function is applicable. This property has nothing to do with the magnitude of the effect size found in the experiment. Hence, it may be concluded that knowing the effect size does not add anything to

Table 3.6   *The logical relations among the instigating observations, abstraction (functional law), research and experimental hypotheses of the generality experiment*

Level of discourse	What is said at the level concerned	
Instigating observations	More rehearsal followed by better recall of words; more rehearsal followed by better singing; more wins after more practice, etc.	
Abstraction	Performance is positively related to the amount of practice for the tasks observed so far.	[P3.6.1]
Generalization (research hypothesis)	Better performance follows from more practice for **all** tasks.	[P3.6.2]
Experimental hypothesis	*If* [P3.6.2], *then* the performance of practised subjects on Task T_1 is better than that of subjects with no practice.	[P3.6.3]
Statistical alternative hypothesis	*If* the consequent of [P3.6.3], *then* $\mathbf{H_1}$.*	[P3.6.4]
Sampling distribution of H_1	*If* H_1, *then* the probability associated with a difference between practised and no-practice subjects as extreme as 1.729 standard error units from an **unknown** mean difference is not known (assuming df = 19).	[P3.6.5]
Sampling distribution of H_0	*If* H_0,† *then* the probability associated with a difference between practised and no-practice subjects as extreme as 1.729 standard error units from a mean difference of zero is .05 in the long run (assuming df = 19).	[P3.6.5′]

*H_1 = mean of practised subjects > mean of subjects with no practice.
†H_0 = mean of practised subjects ≤ mean of subjects with no practice.

knowing whether or not the empirical function can be generalized to a new situation.

In sum, the four types of experiments may be divided into two groups in terms of the ease of distinguishing between the substantive and statistical hypotheses of the experiment. The distinction can be made readily in the case of the theory-corroboration and the clinical experiments; however, it is easy to overlook the distinction in the case of the utilitarian and generality experiments. Be that as it may, that a distinction may be overlooked easily does not mean that the distinction should not be made in meta-theoretical discussions. In fact, much of the misunderstanding of NHSTP may be avoided if the distinction between the substantive and the statistical hypotheses is made in methodological discussion (see Quote [3-1] in Section 3.6.1 for an example).

Not knowing the effect size does not weaken the experimental conclusion in the case of the theory-corroboration, clinical and generality experiments. The effect size may be important in the utilitarian experiment, but only by virtue of some non-statistical criterion. A dilemma arises when the result of a utilitarian experiment is statistically insignificant – say, to abide by the statistical decision or the non-statistical criterion. To opt for the non-statistical criterion is to disregard the lack of statistical significance. However, the better solution is to collect new data with an improved data-collection procedure if there are reasons to question the validity of the statistically non-significant result.

3.7. Empirical Research – Theory Justification

The events encountered in Sections 3.2.1–3.2.4 follow one another temporally in the following order: the to-be-explained phenomenon → theory (substantive hypothesis) → research hypothesis → experimental hypothesis → statistical alternative hypothesis → research data (evidential data). Call this the theory-to-data sequence. Several things may be said about the sequence.

First, the phenomenon → theory (substantive hypothesis) component of the sequence is the theory discovery process. Contrary to what some critics have said (e.g., Schmidt, 1992), empirical research in psychology is seldom carried out to discover or build the theory. Instead, the theory is proposed to account for a phenomenon that exists before any empirical research is carried out. In Popper's (1968a/1959, 1968b/1962) view, there is no prescribed method to arrive at an explanatory theory for the to-be-explained phenomenon. Any speculative account is acceptable. Hence, an explanatory theory is a conjecture. The only constraint is that the speculative account must be consistent with the phenomenon (i.e., the 'prior data-hypothesis' consistency identified in Section 3.2.1).

Second, as the same phenomenon may be explained in numerous ways, numerous explanatory theories are available (hence the idea, 'Reality of multiple explanations', in Section 3.4.2). An important reason why empirical research is carried out is to refine a theory or to choose among the contending alternative explanatory hypotheses. Hence, the theory-corroboration process may also be known as the theory-justification process. In other words, empirical research is carried out to justify theories.

Third, as may be recalled from Section 3.2.4, the statistical alternative hypothesis is the implication of the experimental hypothesis at the statistical level. At the same time, what is said in the experimental hypothesis is determined by the nature of the experimental task. Consequently, the choice of the independent, dependent and control variables of the experiment is determined by the experimental hypothesis in the specific context of the experimental task used.

Fourth, some critics envision a particular kind of empirical research when they argue for the importance of theory discovery. For example, it has been suggested that, to construct theories, one must know some of the basic facts, such as the empirical relations among variables. These relations are the building blocks of theory. (Schmidt, 1992, p. 1177)

[Quote 3-2]

Quote [3-2] is illustrated by Schmidt (1992) with the research question about the relationship between job satisfaction and organizational commitment. What is important to note is that theory in [Quote 3-2] seems to refer to the atheoretical functional relationship between two variables (much like a Skinnerian empirical law), not an explanatory account. Hence, 'theory discovery' may mean no more than establishing the generality of the atheoretical functional relationship so identified. Moreover, questions of the kind raised by Schmidt (1992) are often studied quasi-experimentally (Campbell & Stanley, 1966; Cook & Campbell, 1979), not experimentally. As may be seen in Chapter 4, some criticisms of NHSTP are actually criticisms of using non-experimental methods to corroborate explanatory theories.

3.8.   Summary and Conclusions

The main concern of this chapter is to examine why there is a tendency to conflate the substantive theory with the statistical hypothesis in meta-theoretical discussion. If Tables 3.1, 3.4, 3.5 and 3.6 are taken together, it may be seen that the statistical alternative hypothesis is always the implication of the experimental hypothesis at the statistical level, regardless of the type of experiment involved. It is shown in Table 3.4 that the experimental hypothesis (as well as the substantive hypothesis) of the utilitarian experiment is itself amenable to statistical analysis. For this reason, the distinction between the two hypotheses is blurred in the utilitarian experiment.

At the same time, the utilitarian experiment is used to introduce inferential statistics in statistics textbooks. Consequently, NHSTP is introduced or discussed in the context of a pragmatic research in which no distinction is made among the statistical alternative, experimental, research and substantive hypotheses. Hence, it is understandable that H_1 is confused with the substantive hypothesis. Consequently, H_1 is given the utilitarian import found in some criticisms of NHSTP. This brings about the undesirable consequence of conflating a statistical consideration (whether or not the results are due to chance) with non-statistical issue (e.g., whether or not the outcome is important).

As may be seen, the four types of experiment differ in terms of the ease in distinguishing between the substantive and experimental hypotheses in the experiment. Specifically, the distinctiveness is obvious in the case of the theory-corroboration and clinical experiments, but not in the case of the utilitarian and generality experiments. This distinction renders it possible to explicate the meaning of Meehl's (1967, p. 106) advice that 'the logical structure of the experiment' be improved.

What is shown in Table 3.1 is that, underlying the theory-corroboration experiment, there is a quartet of hypotheses, namely, the substantive, research, experimental and statistical hypotheses. These hypotheses represent successive reasoning steps at the conceptual level, leading from the to-be-studied phenomenon to the statistical hypothesis. However, these steps are obscured when the substantive hypothesis is not an explanatory one or the researcher adopts a utilitarian, rather than an epistemic, goal.

It bears reiterating that H_1 is not the experimental hypothesis, let alone the substantive hypothesis. It is equally important to recall that the substantive hypothesis does not imply H_1 in a simple, straightforward manner. In view of the logical relationships among the quartet of hypotheses, it may also be recalled that the null hypothesis is never used as a categorical proposition in NHSTP. It is not a description of the to-be-studied phenomenon. Instead, H_0 describes the consequence of the data-collection situation if certain assumptions are met. Hence, the 'H_0 is never true' assertion is not a valid criticism of NHSTP.

The chain of deductive inferences, from the substantive hypothesis to the research hypothesis and, finally, to the statistical alternative hypothesis, is a sophisticated one. This sophistication depends on how advanced and well-defined theoretically the postulated hypothetical structures and processes are. It is also determined by the ingenuity of the experimenter in devising the appropriate experimental task, as well as in deducing valid theoretical implications from the substantive hypothesis.

The four types of experiment also differ in terms of the objective of conducting the experiment. It is a pragmatic one in the case of both the utilitarian and the clinical experiments. The concern is an epistemic one for the theory-corroboration and generality experiments. The objective of the theory-corroboration experiment is to test whether or not the to-be-assessed explanatory theory is true. The generality experiment is conducted to test the universality of a functional relationship between two variables.

In other words, the difference between the substantive and the statistical alternative hypotheses is more than just different ways of talking about the same thing. While H_1 is about the effects of some non-chance influences on data at the statistical level, the substantive hypothesis is about the material or efficient cause of the to-be-explained phenomena at the conceptual level.

Statistical significance means simply that chance influences may be discounted because the data are too unlikely to be the result of chance influences. The role of NHSTP is a very limited, albeit important, one. However, to say that something is not due to chance is really not saying much at the theoretical level, particularly when NHSTP says nothing about whether or not an explanatory hypothesis or an empirical generalization receives empirical support. Questions of the exact role of NHSTP and of why the limited role of NHSTP is important will be the topic of discussion in Chapter 4.

Notes

1. A kernel sentence is a simple sentence in the active voice – for example, 'The girls bring a friend.' A kernel sentence represents the minimal structure of a well-formed sentence.

2. The 'should' indicates a logical prescription. This logical prescription is not a forecast of what will happen because events in the world are not determined or caused by an implication of the hypothesis. The experimental implication is a logical prescription in the sense that the hypothesis is not tenable if what is said in the implication is not observed. In other words, it is misleading to talk about 'experimental prediction', the common practice notwithstanding. The better term is 'experimental prescription'.

3. This, of course, is reminiscent of what is said in Section 2.2.4 and of Meehl's (1967, see p. 105, in particular) view that psychological theories are not quantitatively sophisticated enough to prescribe exact magnitudes. However, as may be seen subsequently in Sections 3.4.3 and 3.4.4, the sort of quantitative sophistication required by Meehl (1967) is not relevant to theory corroboration.

4. This argument may also be used to explicate the statement, 'In the scales that psychologists use, zero is usually the only special number' (Frick, 1995, p. 153).

5. It has been shown in Section 2.6.2 that H_0 is never used as a categorical proposition in NHSTP (see also Section 3.5).

6. This point will become important when the issue of induction is discussed in Section 4.5.

4

Evidential Support for a Theory and NHSTP

Testing a statistical hypothesis is not corroborating an explanatory theory at the conceptual level, and NHSTP does not provide evidential support for the theory. The limited, but important, role of NHSTP in empirical research is to supply the minor premiss required to start the chain of embedding conditional syllogisms implicated in theory corroboration. This role is shown in the context of the theory-corroboration experiment. An examination of control shows that it is too much of an oversimplification, if not a misrepresentation, to consider NHSTP an inductive procedure. The concept control is also used to distinguish among different kinds of empirical research. Some criticisms of NHSTP are misdirected because non-statistical functions are attributed to NHSTP in quasi-experimental and non-experimental studies.

4.1. Introduction

A few criticisms of NHSTP have been answered in Chapter 3 in the course of making explicit the logical relationships among the substantive, research, experimental and statistical hypotheses of the theory-corroboration experiment. At the same time, mention is made that (a) NHSTP is not designed to deal with whether or not an explanatory hypothesis or an empirical generalization receives empirical support, and (b) NHSTP has only a limited role in empirical investigation. What is the role of NHSTP? How does NHSTP differ from theory corroboration? Some of the criticisms of NHSTP will assume a different complexion when these questions are answered.

It may be recalled from Chapter 1 that some critics have allowed that their criticisms of NHSTP do not apply to experimental studies (viz., [Q1-2] in Chapter 1; J. Cohen, 1994) or are less readily applicable to experiments using independent variables than those employing subject variables (i.e., [Q1-1] in Chapter 1; Meehl, 1967). How may these qualifications be reconciled with what is shown in Table 3.1 and Tables 3.4–3.6 that (a) H_1 is always the implication of the experimental hypothesis at the statistical level, and (b) H_0 is always the logical complement of H_1, irrespective of experiment type (i.e., theory-corroboration, utilitarian, clinical or generality)?

This consideration will prove helpful in assessing some of the criticisms of NHSTP.

Many of the issues raised so far may be discussed with reference to the theory-corroboration experiment. Hence, the discussion begins with an explication of the rationale of theory corroboration, followed by a discussion of the inductive basis of the experimental design. It will be shown that NHSTP is not an inductive process. A case is made that induction is more than enumeration plus generalization. Moreover, induction serves a purpose considerably more important than effecting generalization. Specifically, induction is used to exclude alternative interpretations of data. This discussion serves as the frame of reference for revisiting other criticisms of NHSTP.

4.2. A Recapitulation

For ease of exposition, it helps to recall the abbreviated version of Savin and Perchonock's (1965) experimental test of the transformational grammar described in Sections 3.2.2 and 3.2.3. After hearing a kernel or negative sentence, subjects were presented with eight extra words on every trial. They were to recall verbatim the sentence and as many as possible of the eight extra words. To facilitate discussion, suppose that the calculated t statistic is 2.05 whose associated probability, p, is .037. At the same time, the critical t is 1.729 for df = 19 (i.e., assume that 20 subjects participate in the repeated-measures one-factor, two-level experiment). It is also helpful to duplicate in Table 4.1 the logical relationships among the phenomenon of linguistic competence, the substantive, research, experimental and statistical hypotheses listed in Table 3.1. The complements of the substantive, research, experimental and statistical hypotheses in Table 3.1 have been omitted in Table 4.1. Added to Table 4.1 are the explications that H_1 represents an exemplification of 'not chance' (see [P4.1.4] and [P4.1.5]) and that H_0 means 'chance' (see *[P4.1.5']*) in view of the discussion in Section 3.3.

4.3. The Rationale of Theory Corroboration

There are two components to the rationale of theory corroboration. The first one is based on the disjunctive relationship between H_0 and H_1. It has been described in Sections 2.2.2 and 2.2.3 how it is justified to make a decision about H_1 by working with H_0. The second component consists of three syllogisms involving the series of conditional propositions [P4.1.1]–[P4.1.5] in Table 4.1. It is in this component that the limited, albeit important, role of NHSTP may be seen. Moreover, together with the design of the experiment, these three conditional syllogisms confer evidential support to the substantive hypothesis.

The series of conditional propositions [P4.1.1]–[P4.1.4] in Table 4.1 forms

Table 4.1   *The logical relations among the phenomenon of linguistic competence, the theory, research hypothesis, experimental hypothesis and statistical hypotheses (alternative and null) in the theory-corroboration experiment*

Level of discourse To-be-explained phenomenon:	What is said at the level concerned The linguistic competence of native speakers of English	
Substantive hypothesis	The linguistic competence of native speakers of English is an analogue of the transformational grammar.	[P4.1.1]
Research hypothesis	**If** [P4.1.1], **then** it is more difficult to process negative sentences than kernel sentences.	[P4.1.2]
Experimental hypothesis	**If** the consequent of [P4.1.2], **then** it is more difficult to remember extra words after a negative sentence than after a kernel sentence.	[P4.1.3]
Statistical alternative hypothesis	**If** the consequent of [P4.1.3], **then** H_1 (as an exemplification of *not chance*).	[P4.1.4]
Sampling distribution of H_1	**If** H_1 (i.e., as an exemplification of *not chance*), **then** the probability associated with a difference between kernel and negative sentences as extreme as 1.729 standard error units from an **unknown** mean difference is not known (df = 19).	[P.4.1.5]
Sampling distribution of H_0	*If H_0, (i.e., chance) then the probability associated with a difference between kernel and negative sentences as extreme as 1.729 standard error units from a mean difference of zero is .05 in the long run (df = 19).*	*[P4.1.5']*

a chain of deductive implications. This aspect of the rationale of theory corroboration may be seen more readily if the reasoning process is represented in the form of three embedding conditional syllogisms (viz., the syllogisms in Roman font, *italics* and **boldface**), as in Table 4.2. The example is one in which a statistically significant result is found.

The three syllogisms are called 'conditional syllogisms' because the major premiss of each of them is a conditional proposition. The innermost conditional syllogism is made up of the three propositions in Roman font, namely, [MAJ-4.2.1], [MIN-4.2.1] and [CON-4.2.1] in Table 4.2. What is said in its minor premiss [MIN-4.2.1], is that H_1 is true. It is, in fact, the conclusion of the disjunctive syllogism discussed in Section 2.2.3.

The second conditional syllogism depicted in Table 4.2 is made up of *[MAJ-4.2.2]*, *[MIN-4.2.2]* and *[CON-4.2.2]*. Its minor premiss, *[MIN-4.2.2]*, is the conclusion of the innermost syllogism (i.e., [CON-4.2.1]). The conclusion of this second conditional syllogism serves as the minor premiss of the last conditional syllogism, which is made up of **[MAJ-4.2.3]**, **[MIN-4.2.3]** and **[CON-4.2.3]**.

It may be seen from Table 4.2 that the minor premiss of the innermost syllogism of the chain of embedding syllogisms is the outcome of NHSTP.

Table 4.2 *The series of three embedding syllogisms (in roman font, italics, and **boldface**, respectively) underlying the theory-corroboration procedure when the null hypothesis is rejected*

Major Premiss 3	If [P4.1.1][1] in Table 4.1, then [P4.2.1].[2]	[MAJ-4.2.3][7]
Major Premiss 2	*If [P4.2.1], then [P4.2.2].[3]*	*[MAJ-4.2.2][6]*
Major Premiss 1	If [P4.2.2], then H_1.[4]	[MAJ-4.2.1][5]
Minor Premiss 1	H_1 is true.	[MIN-4.2.1]
Conclusion 1	Therefore, [P4.2.2] is true in the interim (by virtue of experimental controls).	[CON-4.2.1]
Minor Premiss 2	*[P4.2.2] is true in the interim.*	*[MIN-4.2.2]*
Conclusion 2	*Therefore, [P4.2.1] is true in the interim (by virtue of experimental controls).*	*[CON-4.2.2]*
Minor Premiss 3	**[P4.2.1] is true in the interim.**	**[MIN-4.2.3]**
Conclusion 3	**Therefore, [P4.1.1] in Table 4.1 is true in the interim (by virtue of experimental controls).**	**[CON-4.2.3]**

[1][P4.1.1] in Table 4.1   The linguistic competence of a native speaker of English is an analogue of the transformational grammar.

[2][P4.2.1]   It is more difficult to process negative sentences than kernel sentences (i.e., the consequent of [P4.1.2] in Table 4.1).

[3][P4.2.2]   It is more difficult to remember extra words after a negative sentence than after a kernel sentence (i.e., the consequent of [P4.1.3] in Table 4.1).

[4]H_1 Mean of extra-sentence words recalled after negative sentences < mean of extra-sentence words recalled after kernel sentences.

[5][MAJ-4.2.1]   is [P4.1.4] in Table 4.1.
[6]*[MAJ-4.2.2]*   is [P4.1.3] in Table 4.1.
[7]**[MAJ-4.2.3]**   is [P4.1.2] in Table 4.1.

This is an important role because this minor premiss initiates the chain of deductive inferences. At the same time, this is the only role played by NHSTP. In other words, by itself, the outcome of NHSTP does not confer any evidential support for the substantive hypothesis. Statistical significance only informs the researcher that there is a rational basis for excluding unknown chance factors as the explanation of the data.

In short, it is important to distinguish between theory corroboration and NHSTP. Statistical hypothesis testing, by itself, is not the process of testing conceptual theories. That this distinction is not often made (particularly in statistics textbooks) is the result of failing to see that H_1 is neither the substantive hypothesis itself nor a paraphrase of the substantive hypothesis. The need to distinguish between theory corroboration and NHSTP has been anticipated by a critic of NHSTP, Bakan (1966), as witnessed by [P4-1]:

> after an inference was made concerning a population parameter, one still needed to engage in induction to obtain meaningful scientific propositions. (Bakan, 1966; reprinted in Badia et al., 1970, p. 253)

[P4-1]

[P4-1] is also interesting in that Bakan (1966) had in mind induction whereas a case for deduction is presented in this defence of NHSTP (viz., a series of three embedding conditional syllogisms). As may be seen presently, what is said in [P4-1] is not incompatible with the present argument. The compatibility may be made explicit by considering the asymmetry between *modus tollens* refutation and affirming the consequent confirmation of theories (Meehl, 1967, 1978).

4.4. An Asymmetry and the Issue of Deductive Validity

Table 4.3 shows the series of conditional syllogisms implicated when statistical significance is not found after carrying out NHSTP. As can be seen, the argument implicated in theory corroboration is valid and straightforward when the result is statistically insignificant. Not to reject H_0 is to deny H_1. This amounts to saying that the consequent of [MAJ-4.3.1] in Table 4.3 is false (viz., what is said in [MIN-4.3.1]). Hence, by *modus tollens*, the antecedent of [MAJ-4.3.1] is not true (i.e., [P4.3.2] is not true). This statement becomes the minor premiss of the second syllogism in Table

Table 4.3　*The series of three embedding syllogisms (in* roman font, *italics,* and **boldface**, *respectively) underlying the theory-corroboration procedure when the null hypothesis is not rejected*

Major Premiss 3	If [P4.1.1][1] in Table 4.1, then [P4.3.1].[2]	[MAJ-4.3.3][7]
Major Premiss 2	*If [P4.3.1], then [P4.3.2].*[3]	*[MAJ-4.3.2]*[6]
Major Premiss 1	If [P4.3.2], then H_1.[4]	[MAJ-4.3.1][5]
Minor Premiss 1	H_1 is not true.	[MIN-4.3.1]
Conclusion 1	Therefore, [P4.3.2] is not true.	[CON-4.3.1]
Minor Premiss 2	*[P4.3.2] is not true.*	*[MIN-4.3.2]*
Conclusion 2	*Therefore, [P4.3.1] is not true.*	*[CON-4.3.2]*
Minor Premiss 3	**[P4.3.1] is not true.**	**[MIN-4.3.3]**
Conclusion 3	**Therefore, [P4.1.1] in Table 4.1 is not true.**	**[CON-4.3.3]**

[1][P4.1.1] in Table 4.1　The linguistic competence of a native speaker of English is an analogue of the transformational grammar.
[2][P4.3.1]　It is more difficult to process negative sentences than kernel sentences (i.e., the consequent of [P4.1.2] in Table 4.1).
[3][P4.3.2]　It is more difficult to remember extra words after a negative sentence than after a kernel sentence (i.e., the consequent of [P4.1.3] in Table 4.1).
[4]H_1　Mean of extra-sentence words recalled after negative sentences < mean of extra-sentence words recalled after kernel sentences.
[5][MAJ-4.3.1]　is [P4.1.4] in Table 4.1.
[6]*[MAJ-4.3.2]*　is [P4.1.3] in Table 4.1.
[7][MAJ-4.3.3]　is [P4.1.2] in Table 4.1.

4.3 (i.e., [CON-4.3.1] becomes *[MIN-4.3.2]*). Again, by *modus tollens*, it is concluded that the antecedent of *[MAJ-4.3.2]* is not true (viz., *[P4.3.1]* is not true). *[CON-4.3.2]* is used as **[MIN-4.3.3]** in the third conditional syllogism. This minor premiss denies the consequent of **[MAJ-4.3.3]**. Hence, the third application of *modus tollens* leads to the conclusion that the substantive hypothesis (**[P4.1.1]**) is not true.

However, the argument is strictly problematic in the case of a statistically significant result. As may be seen from Table 4.2, the consequents of the respective major premisses of the three conditional syllogisms are affirmed in succession (i.e., [MAJ-4.2.1], *[MAJ-4.2.2]* and **[MAJ-4.2.3]**). This is strictly invalid because the truth-value of the antecedent of a conditional proposition is indeterminate if the consequent of the conditional propositional is affirmed (Copi, 1982). Specifically, the major premiss, [MAJ-4.2.1], of the innermost syllogism is 'If [P4.2.2], then H_1.' However, knowing that H_1 is true is not informative as to whether or not [P4.2.2] is true. None the less, [P4.2.2] is treated as though it is true in Table 4.2, albeit with the qualification, 'in the interim'. In other words, the actual conclusion of the innermost syllogism, [CON-4.2.1], is 'Therefore, [P4.2.2] is true in the interim.' The 'in the interim' qualification is also found in *[MIN-4.2.2]*, **[MIN-4.2.3]**, *[CON-4.2.2]* and **[CON-4.2.3]** in Table 4.2.

The difficulty with affirming the consequent of a conditional proposition is compounded in the second syllogism when the tentative conclusion of the innermost syllogism is used as the minor premiss. Strictly speaking, the minor premiss of a conditional syllogism should not be a tentative statement. By the same token, the third syllogism is likewise problematic.

In short, the chain of deductive inferences shown in Table 4.3 is valid when a statistically non-significant result is used to initiate the inferential series. However, the validity can be questioned when the inferential series begins with a statistically significant result, as depicted in Table 4.2. This is the asymmetry between *modus tollens* refutation and affirming the consequent confirmation of theories identified by Meehl (1967, 1978). It is for this reason that the various instances of the 'in the interim' qualification in Table 4.2 have to be justified by the stipulation, 'by virtue of experimental controls' (see Section 4.6 below). What are the experimental controls? How do they render it justifiable to accept, in the interim, the antecedent of the conditional proposition when its consequent is affirmed? To answer these questions, it is necessary to digress into a discussion of the logical (inductive) basis of the experimental design.

4.5.   The Logical Structure of the Experimental Design

The importance of the concept control is universally recognized and routinely described in introductory research methods textbooks. For example, it is noted that there is an experimental and a control condition (or group) in the experiment. However, the logical basis of the experiment is seldom, if ever, discussed explicitly. Nor are the three technical meanings of control identified by Boring (1954, 1969) acknowledged in most research methods textbooks. This neglect may be responsible for the fact that control, in the context of research methodology, is sometimes misrepresented to mean shaping the subject's behaviour in the Skinnerian sense, despite Boring's (1954) explicit advice not to do so. This state of affairs, in turn, may be responsible for misunderstanding NHSTP in the form of attributing some of the functions of control to NHSTP.

4.5.1.   Three Meanings of 'Control'

There are three meanings to the term control (Boring, 1954, 1969). They are first explicated with reference to Table 4.4. The issue as to how the three meanings of control feature in the experimental design will be made explicit in Section 4.5.3.

To consider the first meaning of control, ignore the 'Dependent variable' column for the moment, and consider Rows C and E in Table 4.4. They represent the two levels of the independent variable, Sentence-type. Row C defines the control condition, and Row E represents the experimental condition. It may be seen that the two rows are identical, except for the 'Independent Variable' column. This state of affairs shows that the baseline condition (i.e., the one represented by Row C) and the to-be-compared condition (the one depicted in Row E) are identical in all aspects but one. If there is a difference in the number of words recalled between the conditions

Table 4.4   *The inductive basis of the repeated-measures 1-factor, 2-level design (*method of difference)

Independent variable (sentence-type)	Control variables						Extraneous variables				Dependent variable
	C1	C2	C3	C4	C5	C6	*E1*	*E2*	...	*En*	
C Kernel sentence	NI	T	I	R	S	C	*ER*	*IT*	...	*M*	Number of extra words recalled
E Negative sentence	NI	T	I	R	S	C	*ER*	*IT*	...	*M*	Number of extra words recalled

C1 = Normal intonation (NI).
C2 = Task presentation via recorded tape (T).
C3 = Interval between end of sentence and beginning of words (I).
C4 = Rate of word presentation; 3/4 sec per word (R).
C5 = Structure of sentence; 'Animal' subject, present perfect transitive verb (S).
C6 = Fixed categories of words used in 'extra' words (C).
E1 = Extra-curricular reading (ER)
E2 = Individual interests (IT)
En = Kernel and negative sentences randomly mixed (M)

depicted in Rows E and C (i.e., the entries in the 'Dependent variable' column), the difference may be attributed to the difference between the kernel and negative sentences. In this sense, the performance in the control condition is used as a baseline for assessing the performance in the experimental condition. This is the first meaning of *control*, namely, a valid comparison baseline.

The second meaning of control is the constancy of condition requirement. It has two components. The independent variable is Sentence-type whose two levels are Kernel Sentence and Negative Sentence. It is imperative that only simple sentences in the active voice are used in the Kernel Sentence condition, and only simple negative sentences in the active voice are used in the Negative Sentence condition. This is the first component of the constancy of condition requirement. What it means is that the two levels of the independent variable must be used consistently as prescribed in the experimental design.

There are six control variables: C1–C6. They are called 'control variables' because each one of them is used at the same level at both levels of the independent variable. This is what is meant by holding the control variables constant. Take C1 as an example. Variable C1 is intonation. It is represented by the level normal at both the 'Kernel Sentence' and 'Negative Sentence' levels of the independent variable, Sentence-type. This is the second component of the constancy of condition requirement. That is, control variables are those that are held constant at all levels of the independent variable throughout the experiment.

The third meaning of control refers to the provisions in the experiment required for excluding procedural artefacts. This may be explained with reference to the extraneous variables (i.e., variables that are neither independent nor dependent variables nor control variables). In theory, there are numerous extraneous variables. However, only some of them are identified (viz., E1–En). It is assumed in Table 4.4 that these extraneous variables are also controlled for, or held constant, as indicated by the fact that each one of them is represented by the same level (e.g., $E1$ represents extra-curricular reading). This assumption is justified by the fact that the repeated-measures design is used; that is, each subject is tested at both levels of the independent variable. In other words, using the repeated-measures design is a means to prevent any extraneous variable from being a confounding variable (viz., an extraneous variable which varies systematically with the independent variable). That is, using the repeated-measures design is a means of excluding procedural artefact.

As another example of the third meaning of control, consider how the sentences were presented in Savin and Perchonock's (1965) experiment. The kernel and negative sentences were presented in a random order in the course of the experimental session. This procedure was used to exclude the possible order of presentation effects. In other words, randomizing the order of presentation of the levels of the independent variable is another way of eliminating a procedural artefact.

4.5.2. Method of Difference

The schematic representation of the repeated-measures one-factor, two-level design in Table 4.4 is, in fact, the experimental analogue of one of Mill's (1973) canons of induction, namely, method of difference (Boring, 1954, 1969; Chow, 1987a, 1992a; Cohen & Nagel, 1934; Copi, 1982). If one argues in the way Mill (1973) did, one would say that the independent variable is the cause of the result if data are collected in the control and experimental conditions depicted in Table 4.4. However, as pointed out by Cohen and Nagel (1934), Mill's positive interpretation of the logic of method of difference is debatable.

To illustrate Cohen and Nagel's (1934) point, consider the extraneous variable temperament (which is not among E1–En). It is not inconceivable that, as a result of some unknown cultural characteristics, individuals from a particular cultural group react differently to positive (i.e., kernel) and negative sentences, apart from the difference in the processing demand as prescribed by the transformational grammar. Suppose that all subjects come from the said cultural group. In such an event, temperament is a potential confounding variable, despite the fact that every individual acts as the individual's own control. Consequently, it cannot be unambiguously concluded that Sentence-type is the only explanation of the data under such circumstances.

In view of this difficulty, Cohen and Nagel (1934) argue that method of difference is important only in a negative way. Suppose that fewer extra words are recalled under the 'Negative sentence' condition than the 'Kernel sentence' condition. This difference is found even when the two conditions are identical with reference to C1. This means that C1 cannot be the cause of the observed difference between the two levels of the independent variable. By the same token, control variables C2–C6, as well as extraneous variables E1–En, can all be excluded as possible explanations of the data. In other words, the important contribution of Mill's (1973) method of difference is that it makes it possible to exclude alternative interpretations of data, namely, those identified with the control variables or procedures.

4.5.3. Control, Design and Induction

The interrelationships among the functions of control, the design of the experiment and the inductive principle underlying the experiment may now be summarized. Underlying the design of the experiment is an inductive principle which is more sophisticated than the method of 'induction by enumeration' (i.e., enumeration plus generalization) implicit in some criticisms of NHSTP. That is, the function of the inductive rule is not to effect generalization. Induction is used to exclude alternative interpretations of data.

This inductive function is achieved with the three types of control identified by Boring (1954, 1969): (a) a valid comparison baseline for making the comparison, (b) two means of ensuring the constancy of condition (viz., that control variables are held constant and that the levels of the independent variable are used exactly as prescribed in the design), and (c) provisions to exclude procedural artifacts (e.g., using the repeated-measures design, randomizing the order of presentation of the various levels of the independent variable, etc.). The improvement in experimental technique discussed by Meehl (1967) is achieved by fulfilling or refining the three types of control. At the same time, what the inductive principles do is very different from what NHSTP does. That is, it is incorrect to characterize statistics as an inductive process.

4.6.　A Tentative Resolution of the Asymmetry Issue

An attempt may now be made to explain how the asymmetry problem identified by Meehl (1967, 1978) may be dealt with in view of the functions of experimental controls. It may be recalled from Section 3.3 that to reject H_0 is to say that there is a good reason not to explain the experimental data in terms of chance factors. However, saying that something is not due to chance is not saying much. It is necessary to specify which specific non-chance factor is responsible for the experimental outcome. This is another way of reiterating the fact that affirming the consequent of [MAJ-4.2.1] in Table 4.2, by itself, does not logically warrant the acceptance of the antecedent of [MAJ-4.2.1].

Suppose that Savin and Perchonock (1965) used none of the control variables C1–C6 described in Table 4.4, nor the repeated-measures design. While statistical significance renders it possible to say that their result was not due to chance, it is not informative as to which one of the many non-chance factors might be responsible in the absence of any control. Consequently, it would not be valid to say that the antecedent of [MAJ-4.2.1] is true when the consequent of [MAJ-4.2.1] in Table 4.2 is affirmed if no control had been used.

However, variables C1–C6 were used as control variables. Moreover, Savin and Perchonock (1965) did use the repeated-measures design. In satisfying the formal requirement of Mill's (1973) method of difference, they could validly exclude the control and extraneous variables as possible explanations of the data (Cohen & Nagel, 1934). To the extent that all recognized potential confounds have been incorporated into the design as control variables or control procedures, it is reasonable to suggest that the data may reasonably be explained by the manipulation of the independent variable. In other words, it is reasonable to accept the antecedent of [MAJ-4.2.1] in the presence of statistical significance (which rules out chance) when the formal requirement of method of difference is satisfied. This is the basis for the 'by virtue of experimental controls' qualification in Table 4.2.

It is recognized that there is neither any self-evident nor any theoretical justification for the assumption that all potential confounds have been included as control variables. For example, Savin and Perchonock (1965) did not control for sentence-length. Specifically, their kernel and non-kernel sentences differed in length. A more serious difficulty may be their assumption that their sentence-type manipulation provided an opportunity for different types of encoding processing to occur in the course of committing the sentence to memory. However, it is also possible that their manipulation might have affected how kernel and negative sentences were retrieved during the recall phase of the trial.

Recognizing issues like these is partly responsible for the 'in the interim' qualification in Table 4.2. In short, it is accepted that conclusions based on experimental data are inevitably tentative. That is, Meehl's (1967) asymmetry may be given a tentative solution only. However, this tentativeness is not due to the fact that NHSTP is used. Rather, it is the result of the logic involved, namely, that affirming the consequent says nothing about the truth value of the antecedent of a conditional proposition. The tentativeness is also the consequence of the fact that the same phenomenon may be explained in multiple ways. Again, this is not the consequence of using NHSTP. The more important point is that the tentativeness cannot be resolved by appealing to some other statistical indices (e.g., the effect size) or statistical procedure (e.g., meta-analysis). Nor would the tentativeness be reduced by using a more powerful statistical test.

4.7. The Cogency of the Criticisms of NHSTP and Research Methods

It is concluded in Section 4.3 that the role of NHSTP is restricted to providing the minor premiss for the innermost of three embedding conditional syllogisms, regardless of the type of experiment involved. The intriguing question is what may be learned about the criticisms of NHSTP when it is allowed that the criticisms are less cogent when they are directed towards (a) experimental studies (viz., [Q1-2] in Chapter 1; J. Cohen, 1994) or (b) experiments with manipulated independent variables (i.e., [Q1-1] in Chapter 1; Meehl, 1967). For ease of exposition, [Q1-1] and [Q1-2] are duplicated below as [Q4-1] and [Q4-2], respectively:

> Why should NHSTP be more problematic in the case of subject-variable experiments than manipulated-variable experiments?
>
> [Q4-1]

> What renders NHSTP more satisfactory in an experiment than a non-experiment?
>
> [Q4-2]

4.7.1. *Subject Variable and Ambiguity due to Confounding*

To consider [Q4-1], suppose that Psychologist P is interested in whether or not students in the biological sciences differ from their counterparts in the physical sciences in linguistic competence. Psychologist P repeats Savin and

Table 4.5   *Violation of the formal requirement of* method of difference *when a subject variable is used*

Subject variable (Faculty of study)	Control variables						Extraneous variables				Dependent variable
	C1	C2	C3	C4	C5	C6	E1	E2	...	En	
C Biological sciences	NI	T	I	R	S	C	*ER*	*IT'*	...	*M"*	Number of extra words recalled
E Physical sciences	NI	T	I	R	S	C	*ER'*	*IT*	...	*M'*	Number of extra words recalled

C1 = Normal intonation (NI).
C2 = Task presentation via recorded tape (T).
C3 = Interval between end of sentence and beginning of words (I).
C4 = Rate of word presentation; 3/4 sec per word (R).
C5 = Structure of sentence; 'Animal' subject, present perfect transitive verb (S).
C6 = Fixed categories of words used in 'extra' words (C).
E1 = Extra-curricular reading (ER or ER').
E2 = Individual interests (IT or IT').
En = Kernel and negative sentences randomly mixed (M' or M").

Perchonock's (1965) experiment with a one-factor, two-level design in which a subject variable is used. The subject variable is the faculty to which students belong. Its two levels are Biological Sciences and Physical Sciences. For ease of exposition, it is assumed that these two levels are mutually exclusive alternatives. Further, assume that only negative sentences are used.

Table 4.5 is an analogue of Table 4.4 for the subject variable, Faculty of Study. Despite the superficial similarity between the two schemata depicted in Tables 4.4 and 4.5, there are some important differences. First, given the nature of the subject variable, different subjects are necessarily assigned to the *Biological Sciences* and *Physical Sciences* conditions. Moreover, the subject assignment is not (and cannot be) done in a random manner. Consequently, it is no longer valid to assume that variables E1–En are represented by the same level at both levels of Faculty of Study (e.g., Levels ER and ER' of the variable E1 are used at the Biological Sciences and Physical Sciences levels, respectively).

Row C is designated as the control condition, and Row E represents the experimental condition. However, unlike the case with the manipulated variable in Table 4.4, (i.e., Sentence-type), the experimental and control conditions in Table 4.5 differ in many respects (viz., E1–En) in addition to being represented by different levels of the independent variable, Faculty of Study. That is, the formal requirement of method of difference is not met when a subject variable is used. Consequently, it is no longer possible to exclude some extraneous variables. This may explain why Selvin (1957; cited in Oakes, 1986) argued that much of the sociological research cannot be considered experimental studies.

In short, potential confounding variables cannot be excluded when a subject variable is used.[1] With the presence of these potential confounding variables, the asymmetry problem identified by Meehl (1967) is more difficult to overcome. That is, experiments using subject variables are more problematic than experiments using manipulated independent variables because the meaning of the data is more ambiguous. However, this difficulty with using a subject variable is not one that arises from using NHSTP. Instead, it is the result of the fact that the formal requirement of method of difference is violated when a subject variable is used.

4.7.2. Control and Different Types of Research Method

To consider Question [Q4-2], it is necessary to consider the differences among experimental, non-experimental and quasi-experimental research. This can be done by considering the extent to which the formal requirement of an inductive rule is violated (e.g., Mill's method of difference). Specifically, an empirical study may change from an experiment into a quasi-experiment if a recognized control variable cannot be installed in the study. For this reason, some researchers do not consider an empirical study an experiment if no manipulated variable is used in the study. Consequently, the example depicted in Table 4.5 may be treated as an example of a quasi-experiment, in addition to those discussed in great detail by Campbell and Stanley (1966).

Correlational studies are examples of non-experimental studies in which statistics is used. The logic of the correlational study, as well as its difficulties, may be discussed with reference to the three panels in Table 4.6. Depicted in the top panel of the table is the logical structure of Mill's (1973) method of concomitant variation (see also Cohen & Nagel, 1934; Copi, 1982). The schema in the middle panel of the table is the logical structure underlying a correlational study. It depicts a design that may be treated as an extension of Mill's (1973) method of agreement, which is depicted in the bottom panel of the table.

Of interest is the functional relationship between the independent and dependent variables in the top panel, and that between Variables 1 and 2 in the middle panel. Variable X in the top panel is called the independent variable because it is manipulated by the experimenter. However, neither Variable 1 nor Variable 2 is manipulated in the middle panel.

In both the top and middle panels of Table 4.6, X_1, X_2, X_3 and X_4 represent four progressively higher quantitative levels of the variable X. Similarly, Y_1, Y_2, Y_3 and Y_4 represent four progressively higher quantitative levels of the variable Y. That is, the magnitude of Y increases systematically with quantitative increases in X in both cases. Although a positive relationship is depicted in both panels, the argument applies equally well if the relationship is a negative one (e.g., when one increases, the other decreases systematically).

Variables C, K and G are control variables in the top panel because they

Table 4.6   *The schematic representation of Mill's*
Method of Concomitant Variation *(top panel), the*
logical structure of the correlational study (middle
panel), and Mill's Method of Agreement *(bottom*
panel)

Method of Concomitant Variation

Independent variable	Control variables			Dependent variable
X	C	K	G	Y
X_1	C1	K1	G1	Y_1
X_2	C1	K1	G1	Y_2
X_3	C1	K1	G1	Y_3
X_4	C1	K1	G1	Y_4

The correlational study as an extension of *Method of Agreement*

Variable 1	Extraneous variables			Variable 2
X	*C*	*K*	*G*	Y
X_1	*C1*	*K2*	*G6*	Y_1
X_2	*C3*	*K4*	*G1*	Y_2
X_3	*C2*	*K7*	*G3*	Y_3
X_4	*C5*	*K3*	*G2*	Y_4

Method of Agreement

Variable 1	Extraneous variables			Variable 2
X	*C*	*K*	*G*	X
X	*C1*	*K2*	*G6*	Y
X	*C3*	*K4*	*G1*	Y
X	*C2*	*K7*	*G3*	Y
X	*C5*	*K3*	*G2*	Y

assume the same value at all levels of the independent variable (e.g., C1 in all four rows in the case of Variable C). For this reason, these variables can be excluded as explanations of the systematic variation in Y as X varies. (See the reasoning in Section 4.5.2 above.) Consequently, the schema shown in the top panel is the logical structure of an experiment if all recognized control variables have been included in the design. That is, Variables C, K and G in the present example are assumed to exhaust all recognized control variables.

On the other hand, Variables *C*, *K*, and *G* are not control variables in the middle or bottom panel because they are not represented at the same level at the various levels of Variable 1 or 2. For example, *C* has values of *C1*, *C3*, *C2* and *C5* at Levels 1, 2, 3, and 4, respectively, of X. The schema in the middle panel differs from that in the bottom panel in the relationship between

Variables X and Y. Variables X and Y remain constant in the bottom panel, whereas they vary systematically with one another in the middle panel. The bottom panel depicts Mill's (1973) method of agreement. Given the similarity in the absence of any control, the schema shown in the middle panel of Table 4.6 may be characterized as an extension of that depicted in the bottom panel.

In Mill's (1973) view, method of agreement may also be used to establish the causal relationship between X and Y by showing that Y occurs whenever X occurs. To Mill, that Y causes X is demonstrated under such circumstances. However, M.R. Cohen and Nagel (1934) rejected Mill's interpretation by pointing out that X and Y might occur together as a result of a factor other than Y (e.g., a factor not previously recognized or the joint presence of Factors C and K, C and K, K and G, or C, K and G). Worse still, the schema does not permit saying that C, or any other factor, is definitely not the cause. It is possible that any level of C is sufficient to produce X. Hence, M.R. Cohen and Nagel (1934) rejected method of agreement as a satisfactory inductive method.

M.R. Cohen and Nagel's (1934) critique of method of agreement is also applicable to its extension depicted in the middle panel of Table 4.6. None of the variables in the middle panel (viz., C, K and G) can be excluded as an explanation of the variation in Y as a function of changes in X. Hence, the correlational study is not a good theory-corroboration method, even though it may be used to identify the necessary condition for the tenability of the to-be-corroborated hypothesis. The correlational study is a non-experiment because there is no control in the three technical senses of the term.

In short, when all recognized controls are present in an empirical study, it is an experiment (e.g., the schema in Table 4.4). A quasi-experiment is an empirical study in which at least one recognized control variable is absent or cannot be instituted (e.g., the schema in Table 4.5). When none of the recognized controls is found in an empirical study, it is a non-experimental study (e.g., the schema in the middle panel of Table 4.6).

4.7.3. Statistics and Design

What is established in Section 4.7.2 is that the two schemata shown in the top and middle panels of Table 4.6 differ in their ability to exclude alternative interpretations of data. Suppose that two respective sets of data are obtained under the conditions specified by the two schemata. As far as statistics is concerned, researchers may calculate the correlation coefficient between X and Y, or the regression of Y on X, with both sets of data in the same way (assuming that the scale, or level, of measurement is appropriate). Significance tests may be carried out with the resultant coefficients in exactly the same manner. The effect size may be ascertained for both studies. Moreover, the test used in both studies may be equally powerful. Hence, there is no reason why the cogency of the criticisms of NHSTP should differ in the two cases if statistics is *the* consideration.

Data interpretation, however, takes into account the logical structure of the design. For example, it is valid to use the linear regression coefficient in the case of the data from the upper panel because Variable X is manipulated and control variables are present. The propriety of using the linear regression coefficient may be questioned in the case of the data from the schema depicted in the middle panel because there is no manipulated independent variable. (That this nicety is often ignored is a different issue.) As there is no control variable in the schema depicted in the middle panel, only the correlation coefficient is valid for data from the middle panel.

The important point is that this question of validity is not a statistical one. It is one about research design and its logical structure. This point is supported by the fact that J. Cohen (1994) and Meehl (1976) make it clear that their criticisms of NHSTP do not apply to experiments. Meanwhile, experiments are empirical studies whose designs satisfy the formal requirement of Mill's (1973) inductive methods. The moral of the story is that it is necessary to separate statistical issues from issues relating to the design of the empirical study.

4.7.4. Internal Validity

Meehl (1967) reminds researchers that a consequence of improving the experimental techniques is to reduce the standard error of the test statistic. This is the measurement aspect of the experimental techniques. However, conducting an experiment is more than making measurements (Chow, 1987a, 1992a, 1994). In view of what is said about the rationale of theory corroboration in Section 4.5, to conduct an experiment is to collect measurements in two (or more) conditions which are identical in all aspects but one. This suggests that another aspect to the experimental approach is the formal requirement of the design of the experiment. To the extent that this requirement is met, the experiment is said to have inductive conclusion validity, one of the two types of internal validity identified by Cook and Campbell (1979).[2]

It was suggested in Section 4.5.3 that induction and NHSTP are used for different purposes. None the less, in actual practice, there is an intimate relationship between the choice of the inductive method and the statistical procedure. This may be seen from the fact that method of concomitant variation warrants the use of linear regression, whereas the *t* test is used when the inductive method underlying the experimental design is method of difference.

Furthermore, using the same inductive method in different situations may lead to using different versions of the same statistical procedure, for example, the independent-sample and related-sample *t* tests. Either of the two *t* tests is used when method of difference underlies the experimental design. However, the control of extraneous variables is achieved in the independent-sample case by relying on the random assignment of subjects to the two levels of the independent variable. Extraneous variables are controlled in the related-sample case by using the same subjects or matching the subjects in terms of the to-be-controlled variables.

Achieving control by meeting the formal requirement of an inductive rule has two independent consequences. First, the standard error of the differences is reduced (a fact emphasized by Meehl, 1967). Second, it reduces the ambiguity in data interpretation by making it possible to exclude alternative interpretations. These two different aspects of the internal validity of an empirical study are often not made distinct when NHSTP is being discussed. There may be two reasons why this distinction is overlooked. First, both of the two aspects are intimately related to the concept control. Second, the three meanings of control are seldom acknowledged in criticisms of NHSTP.

4.7.5. *Experimental Control and Manipulative Mentality*

The present discussion of control provides a rejoinder to Oakes's (1986) views that (a) experimenters manipulate variables in order to use NHSTP, and (b) using NHSTP induces in researchers a manipulative mentality or a simple-minded trial-and-error approach to experimentation (see [P1-2] in Section 1.10). It may be recalled from Section 3.2.3 that Savin and Perchonock (1965) chose their independent variable and its specific levels with reference to an implication of the transformational grammar. Their choice of the control variables, as represented in Table 4.4, is based on methodological assumptions (e.g., the sentences were read with normal intonation, etc.). These choices are part and parcel of what is called 'experimental manipulation', and they are not carried out in a whimsical manner.

Important to the present discussion, and contrary to Oakes's (1986) characterization in [P1-2] in Chapter 1, the experimenter does not carry out the experimental manipulation because NHSTP is used. Instead, the experimental manipulation serves to satisfy the formal requirement of the inductive rule underlying the design of the experiment. The purpose of achieving the various kinds of control is to exclude alternative interpretations of the data. To the extent the manipulation fails, data interpretation is that much more ambiguous. In other words, far from being an objectionable feature of the experimenter, the 'manipulative mentality' identified by Oakes (1986, p. 48) is what renders an empirical researcher an experimenter (see Section 4.5). Moreover, the experimental manipulation is guided by the theoretical basis of the experiment. There is no inherent reason why using NHSTP would render the experimental manipulation whimsical or simple-minded, Oakes's (1986) misgivings notwithstanding.

4.8. The Intricate Relation between the Two Components of Internal Validity

The conclusion from Section 4.7 is that to improve experimental techniques is (a) to improve the measurements made, and (b) to ensure that the formal requirement of an inductive rule of Mill's (1973) is fulfilled. This is reminiscent of Campbell and Stanley's (1966) and Cook and Campbell's (1979) two types of internal validity, namely, statistical conclusion validity and inductive conclusion validity. While the former is about NHSTP, the latter is about the design of the experiment.

4.8.1. The Oat-Bran Study

The importance, as well as the independence, of the two components of the internal validity of an empirical study may be illustrated with Swain, Rouse, Curley and Sacks's (1990) quasi-experimental choice between two explanations of the putative beneficial effects of eating oat bran. The first one is the Binding to Bile Acids Theory. Being water soluble, oat bran brings about beneficial effects by binding to 'bile acids and [promoting] the excretion of sterols' (Swain et al., 1990, p. 148). On the other hand, not being soluble in water, refined wheat bran does not bind to bile acids. It should not promote the excretion of sterols, according to the Binding to Bile Acids Theory.

The second theory may be called the Displacement of Fatty Foods Theory. Its thesis is that the putative beneficial effects of eating oat bran is not due to any intrinsic value found in oat bran. Rather, Swain et al. (1990) suspected that the beneficial effect of eating oat bran has a more mundane explanation, namely, that someone who eats a lot of oat bran would necessarily eat much less fatty foods. Their aim was to show that anything which reduced the consumption of fatty foods (e.g., refined wheat bran) would be as effective as oat bran in reducing serum cholesterol levels in a properly conducted research.

Using a repeated-measures one-factor, two-level design, Swain et al. (1990) tested the two theories with a four-phase quasi-experiment. The four phases were pretest in Phase I (i.e., the baseline was established), first presentation of food supplement in Phase II (wheat for Group 1, and oat for Group 2), no food-supplement presentation in Phase III, and the presentation of the second food supplement in Phase IV. (Group 1 received oat and Group 2 received wheat.) The independent variable was food supplement whose two levels were oat bran (i.e., the treatment level) and refined wheat bran (i.e., the control level). The amount of serum cholesterol was the dependent variable. There were 20 subjects.

Their control variables were (a) energy and nutrient values of foods in their subjects' 'normal' diet, and (b) the appearance, weight (viz., 100 grams) and taste of food supplements. There were three control procedures, namely, (i) double blind (i.e., neither the investigators nor the subjects were informed of the test condition in Phases 2 and 4), (ii) two balanced orders of food supplement presentation (viz., oat before wheat and wheat before oat), and (iii) random assignment of subjects to the two orders of food-supplement presentation. For every subject, Swain et al. (1990) calculated the difference in the serum cholesterol levels between a food supplement level (oat bran or wheat bran) and a baseline (i.e., when no food supplement was given in Phase I).

Swain et al. (1990) reported that both oat bran and wheat bran reduced the serum cholesterol levels. That is, the supplement-baseline difference was significant and in the expected direction. However, oat bran and wheat bran did not differ in terms of the supplement-baseline difference (i.e., there was no oat–wheat difference). In other words, oat bran and wheat bran were equally effective in reducing the serum cholesterol levels. Swain et al. (1990) concluded that the Displacement of Fatty Foods Theory was supported, but not the Binding to Bile Acids Theory. However, a case may be made that Swain et al.'s (1990) conclusions are debatable because their study lacks inductive conclusion validity, a problem that cannot be rectified by increasing the sample size or increasing the power of the statistical tests used. Nor is the ambiguity reduced when Swain et al.'s (1990) results are assessed in terms of the various confidence-interval estimates available in the report.

4.8.2. *The Lack of Inductive Conclusion Validity*

Underlying Swain et al.'s (1990) study was method of difference. However, the study is a quasi-experiment because some of the recognized controls were absent. Consequently, the formal requirement of method of difference was violated. The most obvious departure is the non-fulfilment of the constancy of conditions requirement. Specifically, Swain et al.'s (1990) subjects were to consume 100 g of oat bran and 100 g of wheat bran in the treatment and control conditions, respectively. However, being left to their own devices, the subjects consumed 87 g and 93 g of oat bran and wheat bran, respectively. That is, Swain et al. (1990) failed to achieve the first component of the constancy of conditions requirement, namely, that the levels of the independent variable should be used consistently as prescribed in the design.

At the same time, the daily intake of energy from the food supplement (a control variable in their study) differed in the oat bran and wheat bran conditions. Specifically, the daily intake was more in the wheat bran condition than in the oat bran condition. Furthermore, there was no guarantee that Swain et al.'s (1990) subjects actually maintained the same 'normal' diet in all four phases. In other words, Swain et al. (1990) also failed to satisfy the second component of the constancy of conditions requirement, namely, that a control variable must be represented at the same level in both the treatment and control conditions.

Swain et al. (1990) also did not handle the control variable diet properly. They used different instruments to measure dietary intake on two different occasions. They began by establishing a participant's dietary-intake baseline with a food-frequency questionnaire in Phase 1. However, they measured the actual amount of food consumed during the food supplement phases (i.e., Phases 2 and 4) by asking participants in their study to keep food-intake records. As acknowledged by them, 'the difference in the reported energy intake may have been the result of the different methods used to assess dietary intake' (Swain et al., 1990, p. 149).

Furthermore, Swain et al. (1990) failed to use a valid comparison baseline when they made some of their comparisons (viz., the first meaning of control). This may be seen from their four-phase procedure. Subjects' blood pressure, body weight and serum cholesterol levels were measured in Phase 1. Subjects in Group 1 were given the treatment level (or the control level for Group 2) of food supplement in Phase 2. Food supplement was withheld for both groups in Phase 3. Subjects in Group 1 were given the control level (or the treatment level for subjects in Group 2) of food supplement in Phase 4. Measures of blood pressure, serum cholesterol levels, food intake and the like in Phase 1 were used as baseline measures for the outcomes of both Phases 2 and 4.

Using Phase 1 data as the baseline for assessing the outcomes of Phase 2 is reasonable. However, the validity of using the data from Phase 1 as the baseline for assessing the data from Phase 4 is debatable. Swain et al. (1990) should have collected new blood pressure, serum cholesterol levels and food intake data in Phase 3, and used them as baseline measures for assessing the outcomes of Phase 4. In other words, Swain et al. violated the valid comparison baseline requirement of control.

It is important to the internal validity of Swain et al.'s (1990) study that their subjects did not know, in Phases 2 and 4, which specific food supplement they received. Otherwise, they might adjust their 'normal' diet accordingly (i.e., the reactivity effect). It is for this reason that Swain et al. used the double-blind procedure. However, 18 of their 20 subjects were aware of receiving the oat bran supplement. This problem was further compounded by the fact that the subjects were asked to weigh themselves at home, as well as to keep a record of what they ate. The reliability of the data is, hence, in doubt because the possibility of reactivity effects cannot be ruled out under these circumstances. In other words, Swain et al. failed to satisfy the third meaning of control.

Swain et al.'s (1990) study may also be faulted on the grounds that they used, as their subjects, individuals who were healthy and sophisticated about diet. Their subjects might have had a healthy diet to begin with. Consequently, Swain et al.'s (1990) result may not be sensitive enough to reveal a real difference between eating oat bran and eating wheat bran. The beneficial effects of consuming oat bran may be minimal if oat bran is added to a healthy diet.

At the same time, knowing the amount of oat bran (or wheat bran) consumed is not the same as knowing the precise amount of water soluble fibres (or water insoluble fibres) if fibre is not the only ingredient in bran. Consequently, it is important to ensure that the amount of water soluble fibres be equal to the amount of water insoluble fibres in the food supplement. Swain et al. reported that their subjects had 38.9 g and 18.49 g of water soluble and water insoluble fibres, respectively. Moreover, a minimum of 60 g to 100 g of water soluble fibres are required to lower the lipid protein level. Therefore Swain et al. had not chosen appropriate levels for their independent variable, in addition to the fact that different amounts were used for the two types of fibre.

This is a serious problem with Swain et al.'s study, despite their excuse:

'It is possible that oat bran in the amount that we used has a small cholesterol-lowering effect (4 per cent) that was not detected in this study. Such a small decrease in response to such a large daily intake of oat bran is unlikely to be important in a practical sense, particularly in view of the uncomfortable gastronin-testional reactions produced'. (Swain et al., 1990, p. 151; my emphasis)

Swain et al. (1990) should also be faulted for confusing conceptual rigour with practical importance. Their professed research objective was to test the tenability of two explanatory hypotheses.

Swain et al.'s choice of dependent variable is also debatable. Recall what is said is the Binding to Bile Acids Theory, namely, that water-soluble fibres bind to 'bile acids and [promoting] the excretion of sterols' (Swain et al., 1990, p. 148). In view of this theoretical assertion, it is curious that Swain et al. did not use the amount of fecal sterols as their dependent variable. This is particularly important because many other factors may reduce the serum cholesterol levels.

The aforementioned reservations about Swain et al.'s (1990) study cannot be alleviated by considering the statistical analyses they have carried out. Swain et al. found a significant supplement–baseline difference, but a non-significant oat–wheat difference in the same repeated-measures study. Under such circumstances, it seems unreasonable to attribute the non-significant oat–wheat difference to insufficient statistical power. At the same time, it also seems unreasonable to attribute the significant supplement–baseline difference to having used too large a sample size. In other words, although Swain et al.'s quasi-experiment may be improved in many ways, none of the need-to-be improved features is determined by, or related to, NHSTP. They are all suggested by the formal requirement of method of difference.

It is often suggested that the confidence-interval estimate is more informative than statistical significance (e.g., Schmidt, 1992). Swain et al. (1990) reported, in addition to the results of various tests of significance, the confidence intervals of various tests. None of confidence-interval estimates is helpful to settle the issues raised about the inductive conclusion validity of their study. Hence, it may be suggested that it is not sufficient for critics to say that knowing the confidence-interval estimates is more important than knowing whether or not there is statistical significance. It is necessary for critics also to explain how knowing the confidence-interval estimates may contribute to theory-corroboration in a way that is not already served by statistical significance.

4.9. Summary and Conclusions

Responsible for Gigerenzer's (1993) impression that NHSTP is treated by psychologists as the *sine qua non* of science is the debatable practice of treating statistical hypothesis testing as though it is theory corroboration. This is one consequence of not distinguishing between the substantive and the statistical hypotheses. It is true that NHSTP is not a means to pass judgement on scientific theories, but not for the reason given by its critics. In actual fact, to corroborate a theory is more than to test a statistical hypothesis.

The evidential support for the theory is secured first by excluding chance influences as an explanation. This is achieved by using NHSTP. The next step is to ascertain the probable non-chance factor by excluding all recognized alternative explanations. This exclusion function is made possible by the inductive principle underlying the design of the experiment. The theory-corroboration process is made up of (a) a binary choice as to whether or not the test statistic has exceeded the critical value, (b) a choice between two conditional syllogisms leading to the choice between an explanation in terms of chance or non-chance influences, (c) a disjunctive syllogism for deciding whether or not the sampling distribution predicated on H_0 is appropriate, and (d) three embedding conditional syllogisms for moving from the statistical hypothesis to the substantive hypothesis via the experimental and research hypotheses. NHSTP serves only to supply the minor premiss for the innermost of three conditional syllogisms.

Critics seem to have ignored the important differences between the two aspects of the internal validity of an empirical study. As has been illustrated with the oat-bran study, regardless of whether or not statistical significance is obtained, the result of an empirical study must be assessed with reference to the inductive logic implicated in the design of the study. Critics are rightly concerned about whether or not the substantive hypothesis is warranted by the data. However, many of their concerns are issues related to the inductive principle underlying the study, not to NHSTP.

Notes

1. The present argument is that, regardless of whether or not there is a difference in the dependent variable between the two levels of a subject variable, the reason for the observed difference is often ambiguous. This thesis is not inconsistent with the view that 'the null hypothesis becomes impossible whenever any of the differences between groups of an experiment is known to have an effect' (Frick, 1995, p. 133). For Frick (1995), the null hypothesis is an assertion that the independent variable does not influence the dependent variable of the experiment.

2. See n. 9 in Chapter 2.

5

Effect Size and Related Issues

Some criticisms of statistical significance based on the putative importance of the effect size are examined, namely, that statistical significance is (a) a matter of sample size, (b) anomalously related to the effect size, (c) wasteful of quantitative information, (d) not informative of the degree of evidential support for the theory, (e) not indicative of the inductive nature of inferential statistics, and (f) uninformative as to the real-life importance of the result. These issues are reconsidered with reference to (a) the difference between the theory-corroboration and utilitarian experiments, and (b) the distinction between the conceptual rigour of theoretical analyses based on empirical data and the practical impact or real-life importance of research results. There are provisions in the research process to guard against cynicism and a cavalier attitude towards research.

5.1. Introduction

The conclusion drawn from Chapters 3 and 4 is that NHSTP is a limited, although important, tool in deciding if it is reasonable to exclude chance influences as an explanation of theory-corroboration experimental data. However, the task of assessing NHSTP is often carried out without distinguishing it from the rest of the research process. Mixing statistical issues with methodological concerns when discussing the inadequacy of NHSTP is misleading enough. The assessment is further complicated by the fact that the statistics–methodology mix is meshed with various non-methodological concerns.

A theme of the present discussion is that many of the criticisms of the concept *statistical significance* should not be directed to NHSTP as a statistical exercise. This is the case because they are issues pertinent to other components of the research process, and should not be used as criticisms of NHSTP. Nowhere is the non-statistical nature of the criticisms of NHSTP more evident than the suggestion that tests of significance be supplemented (if not replaced altogether) by the estimation of the effect size when assessing empirical research.

The argument in favour of using the estimation of the effect size rests in showing that statistical significance provides incomplete information about

Table 5.1 *The putative ambiguity and anomaly of significance tests illustrated with four fictitious studies*

Study	u_E	u_C	Effect size* $d = (u_E - u_C)/\sigma_E$	Statistical test (e.g., t) significant?	df
A	6	5	.1	Yes	22
B	25	24	.1	No	8
C	17	8	.9	No	8
D	8	2	.5	Yes	22

*J. Cohen (1987)

the research result. Statistical significance is faulted for being (a) ambiguous, (b) related anomalously to the size of the effect, (c) not indicative of the degree of evidential support for the hypothesis offered by the data, (d) wasteful because the quantitative information available in the data is not profitably utilized, and (e) not reflective of the inductive nature of inferential statistics. The information about statistical significance is also not conducive to carrying out meta-analysis, a quantitative method of integrating empirical research findings.

It will be argued that these putative shortcomings of NHSTP assume different complexions, depending on whether the research objective is to corroborate explanatory theories, to estimate population parameters or to assess the practical utility of the research manipulations. The assessment criterion is conceptual rigour for theory-corroboration research (Chow, 1991a, 1991b), but 'practical validity' (Rosnow & Rosenthal, 1989) for utilitarian research. To find NHSTP wanting because statistical significance has nothing to contribute to meta-analysis is to assume that meta-analysis is a valid theory-corroboration procedure. Some reservations about meta-analysis will be presented. This discussion begins with a recapitulation of some criticisms of NHSTP based on the effect size.

5.2. The Ambiguity or Anomaly of Statistical Significance

NHSTP has been criticized on the grounds that the outcome of a test of significance is ambiguous or anomalous (called the *ambiguity–anomaly criticism* subsequently). Consider the ambiguity arising from the dependence of statistical significance on the sample size with reference to Studies A, B, C and D in Table 5.1. The effect size is .1 in Studies A and B [viz., $(u_E - u_C)/\sigma_E$].[1] Yet, Study A yields statistical significance, whereas Study B does not. As may be seen from the 'Degrees of Freedom' (df) column in Table 5.1, there are more subjects in Study A than in Study B. It seems that whether or not statistical significance is obtained is a matter of the sample size. Call this critique the sample size-dependence problem. While this critique seems impeccable at the mathematical level, is it reasonable at the empirical level? This is a question that has not been considered in discussion of NHSTP to date.

Studies A and C together (or D and C together) illustrate a situation that is found anomalous by critics. Although the effect size is larger in Study C than in Study A or D, statistical significance is not found in Study C. This state of affairs suggests to critics that statistical significance is not commensurate with the effect size. That is, while a significant result may be accompanied by an effect of a trivial magnitude (e.g., Study A), an effect with a non-trivial magnitude (e.g., Study C) is ignored simply because the analysis is statistically insignificant. This incommensurate significance-size problem critique of NHSTP seems to be predicated on the assumption that the size of the effect is indicative of the degree of evidential support for the hypothesis offered by the data. However, the correctness of this assumption is debatable. For example, what renders a significant effect size of .1 trivial or a non-significant effect size of .9 non-trivial at the theoretical level?

To say that a result is 'statistically significant' is to say that it is an unlikely outcome if it is indeed the result of chance influences (see Section 2.7.2). That is, to say that the result of Study C is insignificant is to deem it explainable in terms of chance influences, the magnitude of its effect size notwithstanding. None the less, some critics recommend taking Study C seriously because the effect is large even though its result is statistically insignificant. Pitting statistical significance against the effect size in this way raises the following chance-versus-size question: Given a statistically insignificant result, what is the minimal magnitude that renders it justifiable to ignore the possibility of chance influences? The further important question is: Why is it justified to ignore the possibility of chance influences? Why is statistics used at all if, in the final analysis, the researcher is prepared to ignore it in favour of some non-statistical criteria, a point already made in Section 3.6.1?

Both Studies A and D yield a significant result. However, the effect size is larger in Study D than in Study A. This information is not utilized when the researcher reports only statistical significance. By the same token, Study C should be treated differently from Study B even though they both yield an insignificant result because the effect is larger in Study C than in Study B. Critics argue that the quantitative information of the effect size should be put to better use. For example, it is suggested that the effect size should be reported so that it can be used in meta-analysis (Glass et al., 1981). This criticism of NHSTP will be referred to as the magnitude-insensitivity problem. To accept this critique in the case of Studies B and C (both yield a non-significant result) is to ignore the possibility of chance influences. This prompts an iteration of the chance-versus-size question. In the case of Studies A and D (both yield a significant result), the criticism is reminiscent of the incommensurate significance-size problem.

The general spirit of the *ambiguity–anomaly* criticism is that the ambiguity and anomaly of statistical significance may be eradicated if researchers are assured that their choice of the sample size is appropriate. Moreover, many critics accept J. Cohen's (1965, 1987, 1992a, 1992b) suggestion that the appropriate sample size is assured if it is determined with reference to statistical power, a topic to be discussed in Chapter 6. Of interest now is the *ambiguity–anomaly* criticism. It may be seen that the criticism is directed to the non-statistical component of empirical research. Specifically, it is used to caution researchers against two probable scenarios.

5.2.1.   Two Implicit Scenarios

Researchers in the first scenario adopt a cynical attitude. They may run a large number of subjects when they wish to ensure a statistically significant result, but only a few subjects when their vested interest is served by a statistically non-significant result. That is, if they wish to, researchers can manipulate statistical significance, by taking advantage of the inverse relation between the test statistic and the sample size.

Researchers are assumed to adopt a cavalier attitude in the second scenario. They choose their sample size by '[appealing] to tradition or precedent, ready availability of data or intuition' (J. Cohen, 1987, p. 52). The main point is that cavalier researchers choose the sample size in a way that is oblivious to considerations of valid research practice. Consequently, their findings are suspect because statistical significance may be the hap-hazard outcome of an opportune choice of the sample size.

To assess the *ambiguity–anomaly* argument, it is necessary to consider the two probable scenarios by examining whether or not (a) researchers accept or reject empirical uncritically results on the *sole* basis of statistical signifi-cance or non-significance, and (b) there are effective non-statistical con-straints on NHSTP users that counteract the *ambiguity–anomaly* argument.

5.2.2.   Safeguards against Cynicism and Cavalier Attitude

It is not clear how true are the cynicism and cavalier attitude envisaged in the *ambiguity–anomaly* criticism. For example, cognitive psychologists accept or reject a finding by reviewing whether or not (a) a proper design has been used (e.g., the repeated-measures or the completely randomized design), (b) subjects are given succinct instructions and sufficient training, (c) proper control variables or procedures are present, and (d) the proper statistical model is used (e.g., the fixed effect or the random effect model of ANOVA; the independent-sample *t* or the dependent-sample *t* test).

In other words, research conclusions (even when they are published) are not accepted or rejected on the *sole* basis of statistical significance or non-significance because psychologists take the internal and external validity of the experiment seriously (Campbell & Stanley, 1966; Cook & Campbell, 1979). In fact, experimental psychologists are so particular about these technical details that some critics of experimental psychology find cognitive psychologists' attention to design and methodological matters too pedantic. For example, cognitive psychologists' concerns with methodology are dismissed as mere 'scientific rhetoric' (Gergen, 1991), and experimental psychologists' non-cynical and non-cavalier attitude towards their methodology has been characterized disapprovingly as 'methodolatry' (Danziger, 1990).

5.2.3. *Constraints arising from Convention and Induction*

The sample size-dependence problem issue is true at the mathematical level. However, it becomes less persuasive when it is recalled that conducting empirical research is not carrying out a mathematical exercise. Take the study of the iconic store as an example (Neisser, 1967). Many theoretical statements about the iconic store are based on data collected from fewer than 15 subjects (sometimes as few as four subjects) who are well trained on the partial-report task[2] (Sperling, 1960). Suppose that, in testing a new implication of a theory of the iconic store with the partial-report task, an experimenter fails to achieve statistical significance when 12 well-trained subjects are used (viz., subjects given two 100-trial training sessions: Chow, 1985). Under such circumstances, to explain the lack of statistical significance by saying that the sample size is small would (and should) not be accepted.

Consider the opposite situation. Suppose that another experimenter uses 30 subjects to test an implication of a theory of the iconic store. The sample size is unusually large for a partial-report study. Further, suppose that the subjects are not given sufficient training on the partial-report task (e.g., subjects were given only 12 training trials on the partial-report task in Merikle's, 1980, study), and the result is statistically significant. Given the fact that the experimental task is unusual and difficult, the inductive conclusion validity of the study becomes questionable when subjects are not given sufficient training on the experimental task, regardless of whether or not the data support the iconic store. (See, e.g., Chow's, 1985, discussion of Merikle's, 1980, conclusion about the iconic store.) Hence, even though the statistical significance suggests that the experimental outcome may not be due to chance, the lack of inductive conclusion validity renders it impossible to isolate the specific non-chance factor responsible for the outcome (see Section 4.5). Hence, the validity of the study should be questioned, the statistical significance notwithstanding.

The point to make is that what is important is that conventions in empirical research include more than just sample size or the α level. The present example shows that the amount of training given to the subjects may be another conventional feature. Moreover, there may be good reasons for adopting a conventional feature. Specifically, extensive training is a necessary conventional feature in partial-report studies because the partial-report task is unusual and difficult. In using a conventional feature (e.g., a conventional sample size), the experimenter is also obliged not to disregard (a) the concomitant conventional features implicated or (b) the convention itself when the result is unfavourable to the hypothesis.[3]

In short, the choice of the sample size in experimental studies is not a haphazard affair. Nor is it as arbitrary as some critics of NHSTP make it out to be. It is true that the choice of the sample size may be guided by the established practice of experimenters in the area. However, the convention itself is based on theoretical and methodological considerations, not simply because this is how something is done in the past. Moreover, the convention also imposes an obligation on the part of the experimenter. The practice of examining reported empirical findings in terms of internal and external validity serves effectively to censure cynicism and the cavalier attitude towards methodological issues, including the choice of the sample size.

5.3. Effect-size Gradation and Degree of Evidential Support

The effect size is a continuous variable (Frick, 1995). This quantitative information is not conveyed by the mere report of statistical significance or non-significance. Hence, critics find reporting merely the statistical significance insufficient. They argue that nothing can be learned about the relationship between two population parameters (e.g., $u_E - u_C$) from the sample statistic (viz., $\overline{X}_E - \overline{X}_C$) when only the statistical significance or non-significance is known. Moreover, it seems intuitively reasonable to assume that the magnitude of the effect size is indicative of the degree of evidential support. That is, may a larger effect size not confer more evidential support for the substantive hypothesis than a smaller effect size?

The answer to this question is not a straightforward one. It depends on the relationships between the substantive and statistical alternative hypotheses discussed in Section 3.2. It helps to recall that, with the exception of the utilitarian experiment (Section 3.6.1), (a) the experimental hypothesis is actually two inferential steps away from the substantive hypothesis, and (b) the statistical alternative hypothesis is not the experimental hypothesis, but its implication at the statistical level.

5.3.1. *The Effect Size in the Utilitarian Experiment*

It is true that the mere knowledge of statistical significance is not informative of the magnitude of the effect.[4] However, the real issue is why it is necessary to have the quantitative information. Behind the magnitude-insensitivity criticism are the assumptions that (i) quantitative information is more important than qualitative information, and (ii) larger effect sizes are more supportive of the to-be-tested hypothesis. These assumptions are reasonable if (a) the experimental hypothesis is a paraphrase of substantive hypothesis, if not the substantive hypothesis itself, (b) the objective of conducting the experiment is explicitly and primarily a utilitarian one, and (c) the to-be-taken practical action is related to the size of the effect.

As these criteria may be met, within limits, in the utilitarian (agricultural model) experiment, the emphasis on the effect size is reasonable when assessing the outcome of the utilitarian experiment under specific conditions. Be that as it may, the relationship between the effect size and the real-life importance of the effect is not necessarily straightforward. For example, the stimulant drug *ritalin* is effective in rendering many hyperactive boys manageable (Swanson & Kinsbourne, 1976). Moreover, it is also true that the boys in question become more manageable as the dosage of the drug increases. However, behavioural manageability was not the concern. They were interested in why there was hyperactivity. For this reason, it is important not to turn a child into a zombie, a state of affairs which might be brought about by using an excessively high dosage of the drug. That is to say, the real-life importance of the research result is not necessarily linearly related to the effect size in the utilitarian experiment. Hence, the 'within limits' qualification is important. At the same time, it shows that 'effect size' is not synonymous with the real-life benefits of the result. However, this important point is not acknowledged in the magnitude-insensitivity criticism of NHSTP.

5.3.2. *The Binary Decision and the Theory-corroboration Experiment*

As may be recalled from Table 3.1, the substantive and the statistical alternative hypotheses are unlike each other in the case of the theory-corroboration experiment (see Section 3.2). Specifically, the substantive hypothesis is a speculative account of the theoretical properties of an unobservable hypothetical mechanism. Meanwhile, the statistical alternative hypothesis is the implication of the experimental hypothesis at the statistical level in the form of a statement about the relationship between the observable independent and dependent variables. It is necessary to emphasize that the hypothetical mechanisms and functions implicated in the substantive hypothesis are qualitatively different from the independent or dependent variables in the theory-corroboration procedure.

In other words, what is said in the experimental hypothesis about the relationship between the independent and dependent variables is not a statement about the hypothetical mechanism. Nor is it a statement about the to-be-explained phenomenon. Instead, the experimental hypothesis is a statement about how the hypothetical mechanism may constrain the said relationship. Furthermore, the objective of conducting the theory-corroboration experiment is solely to test the tenability of the substantive hypothesis. No pragmatic action is prescribed by the outcome of the theory-corroboration experiment.

The apparent attractiveness of using the effect size is that it provides quantitative information. The crucial question becomes whether or not the quantitative information is required for validating the substantive hypothesis. It has been shown (viz., Table 2.1 and Sections 2.2.2 and 2.2.3) that the criterion of acceptance or rejection of the implication of the experimental hypothesis at the statistical level is couched in terms of an ordinal relationship (e.g., $u_{\text{Negative}} < u_{\text{Kernel}}$) or a categorical relationship (e.g., $u_{\text{Negative}} \neq u_{\text{Kernel}}$). That is, the exact magnitude of the difference is *not* prescribed in the experimental hypothesis in either case. Consequently, an ordinal or a categorical relation is all that is required for the purpose of deciding whether or not to exclude chance influences as an explanation. This binary decision is all that is required for starting the chain of three embedding conditional syllogisms implicated in assessing whether or not an explanatory, substantive hypothesis receives support from the data (see Section 4.3).

In short, the theory-corroboration experimenters use an implication of the experimental prescription at the statistical level as a rejection criterion for deciding whether or not to use chance influences as an explanation. This decision requires making a binary decision only (see Section 2.7.3; Tukey, 1960). So long as a numerical difference (with or without specifying the direction of the difference, as the case may be) is established with reference to the appropriate critical value (on the appropriate side of the sampling distribution), the exact magnitude of the difference is immaterial to the argument underlying the theory-corroboration process (see Section 4.3). That is, the magnitude-insensitivity criticism of NHSTP is not applicable to the theory-corroboration experiment because the exact magnitude of the effect plays no role in the rationale of theory corroboration (see also Chow, 1988, 1989).

5.3.3. The Binary Decision of the Clinical Experiment

Consider the task of an examiner responsible for approving or not approving an application for a driver's licence on the basis of a reference score. Once a driver achieves a score equal to, or higher than, the reference score, no further gradation of the applicant's performance is relevant to the decision. As it is a binary decision, the actual difference between the driver's score and the reference score is immaterial once the latter is exceeded by the former. In the case of the clinical experiment (see Table 3.5 and Section 3.6.2), the role of the driving test examiner's reference score is a good description of the role of the critical value of the test statistic for determining statistical significance (e.g., $t_{(df=18)} = 1.734$ at $\alpha = .05$ for the one-tailed test). In other words, the magnitude-insensitivity criticism of NHSTP is also inappropriate for the clinical experiment because the task requires only a binary decision.

Recall from [P1-1] in Section 1.8 Oakes' (1986) disapproval of the fact that NHSTP users are obliged to reject H_0 when $p = .048$, but to retain H_0 when $p = .052$ at $\alpha = .05$ level. His idea is that the absolute difference between .048 and .05 equals that between .052 and .05. The present example of the reference score of the driving test may be used to show that Oakes' displeasure is unwarranted.

Suppose that student-driver A exceeds the reference scores by 5 points and B falls short of the reference score by 5 points. If the driving test examiner were to treat A and B the same, the examiner would be making a mockery of the objectivity of the test. The examiner's integrity might even be called into question under such circumstances. That is, the issue should not be that the examiner is obliged to use the reference score strictly. The question is one about the examiner's integrity, particularly when there is an external pressure on the examiner to ignore the reference score.

The implication of this story on Oakes' (1986) view of the rigidity of the α level or J. Cohen's (1987) disapproval of NHSTP users' rectitude (see note 3) should be obvious. Given the facts that the meaning of α is well-defined (see Section 2.5) and that statistical power is not related to α in the way envisaged in power analysis (a point to be developed in Chapter 6), it is not nonsensical to treat $p = .048$ and $p = .052$ differently. The rigid adherence to this rule speaks ill of neither NHSTP nor its users. Instead, it shows that NHSTP users are to be commended for doing the right thing.

5.3.4. The Ambiguous Cogency of the Effect-size Criticism in the Generality Experiment

Despite the apparent equivalence of the research and experimental hypotheses in the case of the generality experiment, the incommensurate significance-size problem criticism is not as convincing as it may first seem. This is the case because the generality of a hypothesis is determined by the range of situations to which the hypothesis applies (see Section 3.6.3), not the quantitative characteristics of the data. A complication arises because some researchers use empirical generalizations to make pragmatic predictions. Moreover, for positivistic behaviouralists, the shaping of the subject's behaviour may be guided by empirical generalizations. Under such circumstances, the concern with the effect size seems reasonable. However, the quantitative information required is not the effect size itself, but the statistically significant linear regression coefficient. In other words, the cogency of the effect-size criticism becomes mute in the case of the generality experiment. This is the case because the statistical index implicated is the regression coefficient. The point to note is that the criticism based on the incommensurability of statistical significance and the effect size owes its putative cogency to the pragmatic objective of the experimenter (e.g., making predictions), not to the quantitative information conveyed by the effect size.

5.4.   Inductive Inference and Statistical Significance

Inferential statistics is often characterized as an inductive process. Specifically, the sample statistic is the outcome of the enumeration process. It is assumed implicitly in some criticisms of NHSTP that an appropriate statistical rule is then used to effect the generalization in the form of making an assertion about a population parameter on the basis of its corresponding sample statistic. It may be for this reason that some critics of NHSTP consider parameter estimation (viz., the confidence-interval estimate) more important or useful than tests of significance.

However, it has been shown in Section 4.5 that the inductive basis of empirical research is not the induction by enumeration process commonly assumed in the discussion of NHSTP. Instead, Mill's (1973) more sophisticated inductive rules are used to exclude alternative interpretations of data, not to effect generalization. Moreover, the inductive rules underlying the designs of an empirical study and of statistics belong to different domains. In other words, that NHSTP is not reflective of the induction by enumeration process is not a valid criticism of NHSTP. In fact, it is debatable whether or not the induction by enumeration process is as crucial to empirical research as it is commonly envisaged.

In short, as far as the theory-corroboration experiment (see Section 5.3.2) or the clinical experiment (see Section 5.3.3) is concerned, the confidence interval, even when provided, is still used to make a binary decision. For example, statistical significance is determined if the expected population parameter falls outside the 95% confidence interval (see, e.g., Hurlburt, 1993). The actual distance from the interval's boundaries is immaterial. The discussion of Swain et al.'s (1990) oat-bran study in Section 4.8.2 shows that the inductive conclusion validity of the study is not improved by the availability of confidence-interval estimates. This state of affairs is contrary to the position Schmidt (1992, 1994, 1996) advocates.

5.5. The 'Practical Validity' Approach to Assessing Empirical Research

Rosnow and Rosenthal (1989) suggest that knowledge claims made by psychologists have to be justified in terms of the following questions:

Is the treatment effective?

[Q5-1]

How effective is the treatment?

[Q5-2]

How impressive is the treatment?

[Q5-3]

One of the meanings of 'impressiveness' in [Q5-3] is the usefulness of the outcome at the practical level. The concern is the extent to which a research treatment contributes towards the outcome (e.g., an intervention programme such as using aspirin to reduce myocardial infarction, MI). This concern is called 'practical validity' by Rosnow and Rosenthal (1989).

They suggest that the practical validity of a study must be assessed in terms of multiple context-dependent criteria. Statistical significance is, at best, only one of the multiple criteria. At worst, statistical significance does not provide any information about practical validity. It is further suggested that the effect size can supply what statistical significance cannot, namely, the answer to Question [Q5-2]. At the same time, the answer to Question [Q5-3] may be seen more readily if a new measure is derived from the effect size, for example, the 'increase in success rate due to treatment' (Rosnow & Rosenthal, 1989, p. 1279). Furthermore, the new measure may be seen more readily when it is represented with Rosenthal and Rubin's (1979, 1982) 'Binomial Effect Size Display' (BESD).

It is true that statistical significance does not provide answers to Question [Q5-2] or [Q5-3]. At the same time, the meaning of Question [Q5-1] differs for the theory-corroboration and the utilitarian research. However, this state of affairs does not detract from statistical significance being the indication that chance influences may be excluded as the explanation of the result. There is no reason to expect the researcher to answer [Q5-2] or [Q5-3] with NHSTP when NHSTP is not developed for doing such a task. At the same time, it is not clear how the effect size, on its own, can be used to answer [Q5-2] or [Q5-3].

To raise Questions [Q5-1]–[Q5-3] as 'practical validity' questions is to consider some non-statistical issues. The 'practical validity' approach is an argument for adopting real-life consequences as a criterion for assessing knowledge. What needs to be emphasized at this juncture is that questions about the practical validity of the research result are not statistical questions. Nor should practical validity be used as a criticism of NHSTP, because practical validity and statistical significance belong to two different domains. Be that as it may, it is instructive to consider why the practical validity of the research result should not be a criterion for assessing the theory-corroboration experiment.

It is necessary to first distinguish between theoretical and practical knowledge. What is useful at the practical level is not necessarily valid at the conceptual level. The two kinds of knowledge are to be assessed with different criteria. Specifically, theoretical knowledge is to be established by examining the theory–data relation in terms of conceptual rigour (viz., the internal and external validity of the empirical study: Chow, 1991a, 1991b). To assess the practical impact of the research result, on the other hand, it may suffice to ask only Questions [Q5-2] and [Q5-3].

An interesting question is how the concepts *statistical significance* and *effect size* fare in terms of conceptual rigour and practical validity. This may be considered by examining (a) some differences between the utilitarian experiment and the theory-corroboration experiment, and (b) the different sets of questions pertinent to conceptual rigour and of practical validity.

5.6. The Theory-corroboration and Utilitarian Experiments

Some differences between the theory-corroboration and utilitarian (or the agricultural model) experiments have been described in Sections 3.2 and 3.6.1. A notable difference is the relationship between the substantive and

Table 5.2   *Some differences between the agricultural (utilitarian) model and theory-corroboration experiments*

	Example	Agricultural model (Utilitarian) Section 1.2	Theory corroboration Section 3.2
1	Impetus	To solve a practical problem; reflexive of data collection	To explain a phenomenon; independent of data collection
2	Subject matter	The practical problem involving observable events	Unobservable hypothetical entity and its theoretical properties
3	Consequence of research	Take a particular course of action; closure of investigation	Accept tentatively, revise or reject the theory; no closure to the investigation
4	Role of theory	Atheoretical	To-be-tested theory explicitly stated; used to guide experimental design
5	Substantive question	'Is the treatment effective?' 'How effective is the treatment?'	'Why does the phenomenon occur?'
6	Experimental hypothesis	The practical question itself	Qualitatively different from the to-be-assessed substantive hypothesis
7	Experimental manipulation	The to-be-assessed efficient cause itself (see Section 5.7.2)	Different from the to-be-explained phenomenon
8	Dependent measure	The practical problem itself	Different from the to-be-explained phenomenon
9	Statistical significance	To indicate that the explanation of data in terms of chance variations can be ruled out at the α level	
10	Effect	Substantive efficacy (i.e., the consequence of an efficient cause)	The difference between the means of two conditions (i.e., the consequence of a formal or a material cause)
11	Ecological validity	Necessary	Irrelevant, may even be detrimental

experimental hypotheses (see Row 6 of Table 5.2). The experimental hypothesis is qualitatively different from the originating substantive hypothesis in the theory-corroboration experiment. Recall from Savin and Perchonock's (1965) study of linguistic competence that (a) the substantive hypothesis is the psychological reality of the transformational grammar (see Proposition [P3.1.1] in Table 3.1), and (b) the experimental hypothesis is that it is more difficult to remember additional words after remembering a negative sentence than a kernel sentence (see the consequent of Proposition [P3.1.3] in Table 3.1).

The situation is very different in the case of the utilitarian experiment. The experimental hypothesis is a paraphrase of the substantive hypothesis (if not the substantive hypothesis itself) in the utilitarian experiment. Apart from the fact that statistical significance plays the same role in both types of experiment (see Row 9 of Table 5.2), there are many additional differences between the two types of experiment, as may be seen from Table 5.2.

5.6.1. The Utilitarian Experiment

An example of the utilitarian experiment has been described in Section 1.2. The impetus of the utilitarian experiment is to answer a practical, real-life question. The to-be-investigated phenomenon (e.g., the efficacy of Drug D) is not one that exists independently of the research project (viz., administering Drug D). Instead, it is the outcome of the data-collection procedure itself. Hence, the utilitarian experiment may be characterized as a 'reflexive' research exercise. Moreover, how the question is answered determines a subsequent course of action (see Row 3 of Table 5.2). As it is not concerned with explaining why a phenomenon occurs, explanatory theories play no role in setting up the empirical investigation. The experiment is, hence, characterized as 'reflexive' in Row 1 and 'atheoretical' in Row 4 in Table 5.2.

As may be recalled from Section 5.3.1, the size of the effect may be relevant to assessing the utilitarian experiment when (a) both the experimental hypothesis and the dependent variable are related to, if not part and parcel of, the practical question itself (see Row 6 of Table 5.2), and (b) the experimental manipulation is the to-be-assessed efficient cause itself (Row 7 of Table 5.2; see also Section 5.7.2 below). Hence, the utilitarian experiment is assessed in terms of the real-life impact the research manipulation has (see Row 10), as well as of the ecological validity of the experiment (Neisser, 1976, 1988; see Row 11).

It is meaningful, as well as important, to ask how effective the treatment is when the quantitative characteristic of the outcome of the utilitarian experiment determines the subsequent course of action. By the same token, the importance of the effect size diminishes if the subsequent course of action does not bear any functional relationship with the effect size. The effect size also becomes less important if its relationship with the practical or real-life importance of the research result is not a linear one. This point has not been given its due recognition in the 'practical validity' approach. Lastly, the completion of the experiment brings a closure to the investigation (see Row 3 of Table 5.2). This may be responsible for Bakan's (1966) observation that some researchers may not conduct any further investigation after finding a significant result. In sum, Questions [Q5-2] and [Q5-3] are justified in the case of the utilitarian experiment under some circumstances.

5.6.2. The Theory-corroboration Experiment

Section 3.2 gives an example of the theory-corroboration experiment used to examine whether or not there are data that warrant accepting an explanation of a phenomenon (e.g., linguistic competence; see Row 1 of

Table 5.2). Unlike the case of the utilitarian experiment, the to-be-explained phenomenon of the theory-corroboration experiment exists before and independently of the data-collection procedure. Hence, the 'independent' characterization is used in Row 1 of Table 5.2. The explanatory account is made up of unobservable, hypothetical mechanisms and functions (e.g., a linguistic processor, the recursive applications of the transformational grammar, etc.; see Row 2 of Table 5.2), to which well-defined theoretical properties are attributed (e.g., the mental representation of the transformational rules).

It is not meaningful to ask, 'How effective is the treatment?' in the theory-corroboration experiment (see Row 5 of Table 5.2) because (a) its substantive 'why' question is not answerable in quantitative terms, (b) the substantive question is not answered directly, but via a different experimental hypothesis, and (c) both the experimental manipulation and the dependent variable are different from the to-be-explained phenomenon (i.e., Rows 7 and 8 of Table 5.2), as a result of Point (b). Contrary to a commonly held assumption, ecological validity is actually detrimental to the validity of the study (Chow, 1987d; Mook, 1983; see also Section 7.8). The conclusion of the study is to accept tentatively, revise or reject the theory (viz., Row 3 of Table 5.2). The completion of the experiment, however, does not mean the closure of the investigation because there are potentially many other alternative explanatory accounts at the conceptual level, as explained in Sections 4.4 and 4.6 (see also Section 6.10.4).

5.7. Questions about Effect, Utility and Validity

In view of the many differences shown in Table 5.2 between the utilitarian and theory-corroboration experiments, it is not surprising that different questions are asked about the two kinds of experiment, as may be seen from Table 5.3. This state of affairs may be traced to the connotative meaning of 'effective' in Question [Q5-1], namely, 'Is Treatment T effective?' Question [Q5-1] is labelled [PV-1] and [CR-1], respectively, for the utilitarian and theory-corroboration experiments.

At the level of statistical discourse, 'effect' refers to the difference between the means of two (or more) levels of an independent variable, r, r^2, or several other indices (J. Cohen, 1965; Rosnow & Rosenthal, 1989). 'Effective', at this technical level, refers to a difference being too extreme (and, hence, deemed too unlikely) to be explained by chance influences on the data-collection procedure. There is no intimation of causal efficacy in this technical meaning in any of the indices. As has been argued in Sections 5.3.2 and 5.3.3 above, questions about 'effectiveness' in its technical sense are binary questions. For this reason, Question [Q5-1] is labelled [CR-1] in the 'Theory-corroboration research' column in Table 5.3.

The term 'effect', none the less, suggests efficacy of some sort. The sort of efficacy implied differs in the theory-corroboration and utilitarian

Table 5.3 *Different sets of research questions pertinent to practical validity (PV) and conceptual rigour (CR) for the utilitarian and theory-corroboration experiments, respectively*

Practical validity concerns (Utilitarian research) The independent variable is the efficient cause		Conceptual rigour concerns (Theory-corroboration research) The independent variable is the material or formal cause	
[PV-1]	Is Treatment T effective? [Q5-1]	[CR-1]	Is Treatment T effective? [Q5-1]
[PV-2]	How effective is Treatment T? [Q5-2]	[CR-2]	Is the independent variable a valid choice?
[PV-3]	How impressive is Treatment T? [Q5-3]	[CR-3]	Do the data warrant the acceptance of Theory K which underlines the choice of the dependent variable?
[PV-4]	Is Treatment T important?	[CR-4]	Is the implementation of the independent variable valid?
		[CR-5]	Does the study have hypothesis validity?

perspectives. This may be seen from the fact that the independent variable plays different causal roles in the theory-corroboration and utilitarian experiments. Specifically, the independent variable is a material or formal cause in the former, but an efficient cause in the latter.[5] Hence, Question [Q5-1] is labelled [PV-1] in the 'Utilitarian research' column in Table 5.3.

As an efficient cause is the agent whose activity produces the effect, it would make sense to ask Questions [PV-2]–[PV-4] in the case of the utilitarian research. That is, practical validity questions are important questions when an efficient cause is implicated. The situation is quite different in the case of a material or formal cause.

Questions about the material cause of X are questions about the nature of X (or about the material of which X is made). To ask questions about the formal cause of X is to ascertain the pattern assumed by the material in X. For example, what distinguishes a bronze statue from any shapeless piece of bronze is its form, the likeness of a boy. In either case, the questions are qualitative, not quantitative, questions. Hence, Questions [CR-2]–[CR-5] are the appropriate questions, not Questions [PV-2]–[PV-4], when a formal or a material cause is involved.

5.7.1. The Independent Variable as the Formal or Material Cause

The independent variable in the linguistic competence example in Section 3.2 is Sentence-type whose two levels are kernel and negative sentences. The dependent variable is the number of extra words recalled after the verbatim recall of the sentence. In theoretical terms, the hypothetical linguistic processor is given a task to do (to retrieve as many words as possible from a list of eight extra words) in two different conditions (i.e., to retrieve a kernel sentence in one condition and a negative sentence in the other condition before retrieving the extra words).

The two test conditions are set by the two levels of the independent variable. Neither of these two levels is the to-be-studied linguistic processor. Nor are the two levels the theoretical properties of the hypothetical linguistic processor. Hence, the independent variable may arguably be characterized as a means to reveal a property of the hypothetical linguistic processor. The independent variable may hence be characterized as a material cause if the theoretical emphasis is on the nature of the mental apparatus (i.e., of what the mental apparatus is made). The dependent variable is the outcome of the behaviour of the hypothetical linguistic processor. However, it may be noted that retrieving the extra words is not what linguistic competence is. Consequently, the independent variable may alternatively be characterized as a formal cause if the research interest is about the behaviour of the hypothetical mechanisms (e.g., the constraint to which the linguistic processor is subjected).

In either case (i.e., the independent variable being the material or formal cause), the independent variable is not an efficient cause because the grammar of the sentence is not the agent that produces the observed outcome, and the dependent variable is not the to-be-studied phenomenon. It is for this reason that ecological validity is irrelevant, as well as detrimental, to theory-corroboration experimentation (Chow, 1987d; Mook, 1983; see also Section 7.8). In other words, it is easy to maintain the technical meaning of 'effective', namely, a difference being too extreme to be explained by chance influences.

More important, the binary nature of Question [CR-1] (in its exclusively technical sense) precludes questions like Questions [PV-2]–[PV-4] in the 'Utilitarian Research' column of Table 5.3. Instead, Question [CR-1] leads explicitly to further questions about the internal and external validity of the experiment. As none of Questions [CR-2]–[CR-5] in the 'Theory-corroboration Research' column of Table 5.3 requires quantitative information, it is not necessary to ascertain the effect size.

A consideration relevant to Question [CR-2] of Table 5.3 ('Is the independent variable a valid choice?') is the choice of the levels of the independent variable. For example, an important independent variable for the study of the iconic store is the inter-stimulus interval (ISI), that is, the interval between the offset of the stimulus and the onset of the probe tone (see n. 2 above). As it is generally accepted that the iconic information remains useful for up to 500 msec only, some typical ISI values used to map the decay of the iconic information are 0, 50, 100, 250, 500 and 1000 msec. Hence, it would be generally inappropriate, or invalid, to use ISI values of 1500, 1750, 2000 and 2500 msec for such a purpose. The point of interest is that Question [CR-2] has nothing to do with the quantitative information conveyed by the effect size.

The issue germane to Question [CR-3] of Table 5.3 ('Do the data warrant the acceptance of Theory K which underlines the choice of the dependent variable?') may be illustrated with Swain et al.'s (1990) study described in Section 4.8. Suppose that Theory K is Swain et al.'s (1990) Binding to Bile Acids Theory, which says that water soluble fibres bind to bile acids and, thus, promote the excretion of sterols. Yet Swain et al. used the serum cholesterol levels, instead of the amount of fecal sterols, as their dependent variable. What is learned about the serum cholesterol levels is not informative as to the amount of fecal sterols. In other words, Swain et al.'s choice of their dependent variable is inappropriate. Consequently, the absence of a difference between the oat and wheat conditions in terms of a dependent variable not stipulated in the experimental hypothesis does not mean that the Binding to Bile Acids Theory is untenable. On the other hand, even if there were a difference between the serum cholesterol levels under the oat bran and wheat bran conditions, the data would still be uninformative as to the tenability of the Binding to Bile Acids Theory because serum cholesterol levels may be reduced by other factors.

The discussion of the lack of inductive conclusion validity in Swain et al.'s (1990) study in Section 4.8.3 is also a good illustration of what Question [CR-4] in Table 5.3 is all about. Specifically, Swain et al. did not ensure that their subjects actually consumed the amount of oat bran or wheat bran prescribed in the design of the study. That is, they failed to observe the constancy of condition requirement.

Question [CR-5] in Table 5.3 arises because, logically speaking, it is not possible to prove any theory with absolute certainty. The problem arises because of the facts that (a) the theory-corroboration procedure starts with a conditional proposition like Proposition [P4.1.2] in Table 4.1, and (b) affirming the consequent of a conditional proposition says nothing, with certainty, about the antecedent of the conditional proposition (see Sections 4.3 and 4.4). The best one can do is to exclude as many alternative explanatory theories as possible.

5.7.2. The Independent Variable as the Efficient Cause

The situation is different in the utilitarian experiment. Recall from Section 3.6.1 that, in the case of the utilitarian experiment, the experimental hypothesis and experimental manipulation are the substantive hypothesis and substantive manipulation, respectively. Consequently, the research manipulation is the efficient cause. Hence, 'effect' refers to the consequence of an efficacious agent. By the same token, 'effective' in the practical validity perspective refers to being substantively efficacious.

As an illustration, the independent variable in the example in Section 1.2 is Method of Teaching. Both of its levels, namely, the New Method (E) and the Orthodox Method (C), are courses of action that could shape the subjects' behaviour. For this reason, in addition to its technical meaning, the term 'effective' may also be attributed the meaning of being substantively

Table 5.4   *The binomial effect-size display (BESD) for the effect of aspirin on myocardial infarction (MI) for all subjects (adapted from Rosenthal, 1989)*

Effect size, r	Treatment condition	% MI absent	% MI present	Change in success rate (%)
.034	Aspirin	51.7	48.3	3.4
	Placebo	48.3	51.7	

efficacious. This is what renders meaningful questions about the effect size, the impressiveness and the practical importance of the empirical study (i.e., Questions [PV-2], [PV-3] and [PV-4], respectively, in Table 5.3). An example may be Rosnow and Rosenthal's (1989) 'practical validity' approach.

5.7.2.1.   Practical   Validity   and   the   Binomial   Effect-size   Display (BESD)   A large number of physicians were divided into two groups in a five-year study (Steering Committee of the Physicians' Health Study Research Group, 1988, cited in Rosnow & Rosenthal, 1989). One group received an ordinary aspirin tablet every other day; the other group received a placebo tablet under the same regime. This intervention procedure is said to have practical validity if the procedure brings about a change in the preferred, beneficial direction (viz., fewer occurrences of myocardial infarction, MI).

Using part of the original data, Rosnow and Rosenthal (1989) illustrated the 'practical validity' approach by considering the number of MI incidents in the Aspirin and Placebo groups. They used Rosenthal and Rubin's (1979, 1982) 'binomial effect-size display' (BESD) to show the practical importance of the effect of aspirin on minimizing the occurrence of heart attack. An example of BESD may be found in Table 5.4. The BESD proper is the 2 by 2 contingency table made up of two levels of MI (viz., the %MI Absent and %MI Present columns) and two levels of medication (viz., the Aspirin and Placebo rows). The correlation coefficient, r, is the index of effect size. How the effect size is related to its BESD, as well as how the 'Change in success rate' is established as the index of practical validity, may be illustrated as follows:

(1)   The effect size is represented by the Pearson correlation coefficient between the independent and dependent variables (viz., $r = .034$).

(2)   The index, success rate, for the Aspirin group is $.5 + r/2$ and $.5 - r/2$ for the Placebo group (Rosenthal & Rubin, 1979, 1982). By 'success' is meant the absence of myocardial infarction (MI) in Rosnow and Rosenthal's (1989) example. In general, 'success' refers to the category or state of affairs consistent with the pragmatic substantive hypothesis.

The Aspirin success rate is 51.7%, and the Placebo success rate is 48.3% (see the Aspirin and Placebo rows, respectively, in the '%MI Absent' column in Table 5.4).

(3) The practical importance is indicated by the 'Change in success rate', namely, the difference in success rate between the Aspirin and Placebo conditions. It is 3.4% in this example (viz., 51.7% minus 48.3%).

5.7.2.2. Practical Validity and Extra-statistical Considerations   In other words, the index of practical validity is simply the correlation coefficient expressed in percentage form. Given a change of this magnitude (viz., 3.4%), 'the implications are *far from unimpressive*' (Rosnow & Rosenthal, 1989, p. 1279; emphasis added). This is the extent of the explication of the relationship between the effect size of the intervention programme and the real-world importance of the effect in the 'practical validity' approach. The BESD is said to be useful because it is an 'intuitively appealing general-purpose effect-size display . . . [that is] easily understood by researchers, students, and lay persons' (Rosenthal, 1983, p. 11). However, intuitive attractiveness or intuitive reasonableness is not justification. For example, why a change of 3.4% is 'far from unimpressive' is not self-evident.

Another reason why the intuitive appeal of BESD is debatable may be illustrated with the derivation of the success rate. The equation is based on the assumption that 'the mean outcome in one group [e.g., the experimental or Aspirin] is the same amount above .5 as the mean outcome in the other group [e.g., the control or Placebo] is below .5' in the case of a dichotomous outcome variable (Rosenthal & Rubin, 1982, p. 168). When the outcome variable is a continuous one, it is assumed that the outcome variable has the same variance in both the experimental and control groups. It is claimed that the validity of BESD is not affected when a continuous outcome variable is dichotomized under such circumstances (see Table 4 in Rosenthal & Rubin, 1982, p. 168).

What is not clear, and is not explained by Rosenthal and Rubin (1982), is why '.5' is chosen. The reason for this reservation about '.5' may be seen by considering an important issue relating to how a continuous variable is dichotomized. Where should the dichotomy criterion be located on the continuous range of outcome values? The assumption underlying the equation, 'Success rate $= 0.5 \pm r/2$', suggests that the mean or median of the distribution of the outcome variable may be the location of the criterion. This seems to be a problematic assumption.

Take mathematical skills as the outcome variable. Suppose that the empirical question is whether or not the new method of teaching would turn students into superior mathematicians. Under such circumstances, it does not seem reasonable to use the mean or median score of a test as the criterion for effecting the dichotomy. As an alternative, the score at the 75th or the 90th centile rank of the test may be more reasonable in a different situation. Again, 'being reasonable' is not a justification. That is, there is still the remaining issue of how the criterion of dichotomy may be justified. In sum, the rationale of BESD is not self-evident, the claim of intuitive appeal notwithstanding.

Suppose that a new way of making films is assessed, and that a favourable change of 3.4% in the audience's attitude is found when the outcome variable *Attitude towards Films* is dichotomized into Favourable and Un-favourable. Is a change of this magnitude impressive? For a film critic, such a magnitude may be unimpressive. However, the assessment is different in the case of the outcome variable myocardial infarction. The occurrence of MI is serious, even fatal. It is not unreasonable to subscribe to the view that an intervention procedure is worthwhile even if it saves only one life. A change of 3.4% is understandably impressive in such a context.

The myocardial infarction and film examples show that the real-life importance of a research intervention is not determined by the magnitude of the change itself. That is, the issue is really not the numerical value of Rosnow and Rosenthal's (1989) 'change in success rate'. Rather, this assessment is based on the evaluation of the nature of the outcome measure itself (e.g., fatality versus the critical success of films). This seems to be contrary to the objective of the BESD exercise, namely, to provide a numerical index for the real-world importance of the research result.

Suppose that the 3.4% favourable change in attitude towards films is assessed by a film producer. When this percentage is translated into monetary figures, the profits may be substantial enough to be considered 'far from unimpressive'. That is to say, the assessment of real-world importance is also dependent on the vested interests of consumers of the research result (viz., the film critic versus the film producer). It depends on neither the magnitude of the effect size nor the nature of the outcome.

In sum, it has been shown that the impressiveness of the pragmatic consequence of research results is in the eye of the beholder. How impress-ive the research outcome is may depend on the nature of the outcome or the vested interests of the beholder, not necessarily on the magnitude of the to-be-measured effect itself. This suggests that statistical conclusion validity and practical validity belong to two unrelated or independent domains. It is instructive to look more closely at the independence in question.

5.7.3. *Statistical Conclusion Validity and Practical Validity*

NHSTP is found wanting because statistical significance is not informative as to the importance of the result at the practical level. Moreover, it is said, 'neither experienced behavioral researchers nor experienced statisticians had a good intuitive feel for the practical meaning of such common effect-size estimators as r^2, omega^2, and similar estimates' (Rosenthal, 1983, p. 11). Under such circumstances, a casual reading of the 'practical validity' approach may lead one to conclude that the effect size (as measured by r or expressed with BESD) can supply the information about the practical meaning or importance of research results. However, it has just been shown that a 'change in success rate' value owes its real-life meaning, as well as its intuitive appeal, to either the nature of the to-be-measured intervention procedure or the researcher's vested interests, not to the numerical value of the effect-size index itself.

Perhaps it is inappropriate to expect statistics users (however experienced) to have 'a good intuitive feel' for the practical validity of an empirical study on the basis of a statistical index for the simple reason that this is not what statistics, *qua* a mathematical research tool, is capable of doing. It may be proper to restrict the role of statistics to giving a succinct description of data or deciding whether or not chance influences may be excluded as the explanation of the result.

It has been shown that even the magnitude of the effect of the research manipulation of the utilitarian experiment cannot be used to answer Questions [PV-3] and [PV-4] in Table 5.3 because the real-life meaning of a particular effect size depends on some extra-statistical criteria. Consequently, even if the size of the effect is reported, it is still not clear how 'impressive' or 'important' the result is in the absence of some extra-statistics criteria. It is true that statistical significance is also not helpful in answering these questions. Statistical significance is, none the less, used to answer Question [CR-1] in its technical sense (see Section 2.7.3). Questions [PV-3] and [PV-4] in Table 5.3 are not appropriate for the theory-corroboration experiment because they implicate evaluation criteria unrelated to conceptual rigour.

In short, the chance-versus-size question identified in Section 5.2 arises because of the conflict between the technical and connotative meanings of 'effect' or 'effective'. The decision is to ignore an insignificant effect, whatever its magnitude, if one adheres solely to the technical meaning of 'effect'. This becomes more difficult when assessing a utilitarian study because (a) the experimental hypothesis is indistinguishable from the substantive hypothesis, (b) the research objective is a pragmatic one, (c) the outcome variable itself has real-life consequences, and (d) it may not be possible for the researcher to ignore the vested interests of some parties connected with the research for reasons unrelated to methodological or conceptual rigour.

However, it is precisely because of these difficulties that researchers should avoid being misled by the connotative meaning of 'effect' or 'effective'. It does not belong to the statistical or methodological discourse. The validity or usefulness of statistical significance or effect size is not related to the practical importance of the research result. A separate decision implicating different criteria is required for determining the practical validity of the result. Rosnow and Rosenthal's (1989) contribution is the introduction of the concept practical validity. The present argument is that the practical validity of an empirical study is not a statistical issue. Hence, NHSTP should not be evaluated in terms of practical validity. Moreover, there are reservations about the propriety of using the effect size as an index of practical validity.

5.8. Meta-analysis Revisited

One of the arguments in favour of reporting the effect size is that this quantitative information may be used in meta-analysis, a numerical method of integrating research results. Such a need arises in the following situation. Suppose that some researchers found psychotherapy beneficial to patients, whereas other researchers argued that psychotherapy is ineffective. This is a confusing situation for policy makers or fund granting agencies. A method is needed to streamline the information for the policy makers.

The meta-analytic approach is such an approach. It was originally proposed as a means to provide a summary numerical statement for policy makers about the efficacy of an intervention programme (e.g., whether or not psychotherapy is beneficial to patients) in view of the conflicting statements about its usefulness. That is, meta-analysis was originally developed as a means to influence decision makers in some bureaucracy. This motivation renders it understandable (but not justifiable) why conceptual rigour or research quality is not deemed important in meta-analysis. For example, it has been claimed that the meta-analytic method makes it possible to achieve 'Garbage-in – information out' (Glass & Kliegl, 1983, p. 37). However, nothing in the meta-analytic process shows how a valid conclusion may be drawn from aggregating data from a set of studies, some of which are invalid studies. It is not clear how it is possible to come up with valid information if pieces of invalid information are integrated in the meta-analytic manner.

It is necessary to discuss meta-analysis because the meta-analytic approach is more recently used as a means to validate theoretical concepts (e.g. Cooper, 1979; Cooper & Rosenthal, 1980; Harris & Rosenthal, 1985; Schmidt, 1992, 1994, 1996). Hence, it is important to consider whether or not meta-analysis is a valid theory-corroboration tool. An important consideration in making knowledge claims is whether or not research data warrant accepting theoretical statements. For this end, conceptual rigour is important when considering the theory–data relationship. However, conceptual rigour is not an issue in meta-analysis, as may be seen from the fact that some meta-analysts deny the distinction between good and poor research (Glass, 1976, 1978).

5.8.1. What Is Meta-analysis?

That 'meta-analysis' is the analysis of analyses may be explained by considering the fictitious data shown in the 'p of Test statistic' and 'Effect size' columns in Table 5.5. Depicted in the table are the p values of the test statistic used and the estimates of the effect size of 12 studies of a particular phenomenon. It is argued by meta-analysts that these p values or effect-size estimates may be used to obtain a combined Z or the combined effect size. Combining these values of p or effect sizes is a form of data accumulation to meta-analysts.

Table 5.5   *The incommensurability difficulty of meta-analysis illustrated with fictitious 'raw data'*

Study	p of test statistic	Effect size[1]	Independent variable	Property or function of the iconic store studied
1	.021*	.7	ISI[2]	Rate of decay
2	.001*	·3	Type of task	Relatively large storage capacity
3	.050*	.5	Number of concurrent tasks	Independence from the short-term store
4	.110	.11	What to recall	Independence of location and identity information
5	.068	.17	Stimulus material	Non-associative information
6	.049*	.4	Type of task	Visible persistence
7	.02*	1.5	Type of material	Unprocessed information
8	.046*	1	Time of probe presentation	Select before processing
9	.070	.06	ISI	Visible persistence
10	.04*	.18	Stimulus duration	Information registration rate
11	.038*	.2	Stimulus duration within a fixed SOA[3]	Information registration rate
12	.066	.29	Type of material	No identity information
	Combined Z	Combined effect size		

*denoted significance at the .05 level.

[1]J. Cohen's (1987) $d = (u_E - u_C)/\sigma_E$.

[2]'ISI' refers to the inter-stimulus interval, the interval between the offset of the stimulus and the onset of the partial-report tone (see n. 2 in this chapter).

[3]'SOA' refers to stimulus-onset asynchrony, the interval between the onset of the stimulus and the onset of the mask.

A statement about the overall statistical significance is then made on the basis of the combined Z, called 'combined significance level' (Harris & Rosenthal, 1985). That is, the p, or effect-size, values from individual studies are treated as raw data (hence 'raw data' in the table's title) and subjected to statistical analysis at a higher level of abstraction. This more abstract analysis is called meta-analysis or the 'analysis of analyses' (Cooper, 1979; Cooper & Rosenthal, 1980; Glass, 1976, 1978; Glass & Kliegl, 1983; Glass et al., 1981; Harris & Rosenthal, 1985; Schmidt, 1992).

5.8.2.   The Meta-analytic Assumptions about p and Effect Size Reconsidered

The assumption of using the combined Z to corroborate theories is that p says something about the substantive hypothesis. There are three necessary conditions for this assumption to be true in the case of theory corroboration. The first one is that NHSTP is the theory-corroboration procedure. However, this has been shown to be not the case in Section 4.3.

The second necessary condition is that the experimental hypothesis, if not the statistical hypothesis, is equivalent to the substantive hypothesis. While this is true of the utilitarian empirical research (viz., the kind of empirical studies meta-analysts were originally concerned with), it is not true of theory-corroboration experiment (see Section 3.2).

The third necessary condition is that p is a probability that the null hypothesis is true. However, it is shown in Sections 2.8.2 and 2.8.4 that p is nothing of the sort. Rather, it is the probability of obtaining a test statistic as extreme as, or more extreme than, the obtained test statistic *contingent on* chance influences. The p value does not confer any evidential support for the substantive hypothesis. In sum, none of the three conditions necessary for the validity of the meta-analytic approach is met.

It is assumed in meta-analysis that the effect size is relevant to the issue of whether or not the research data warrant accepting the hypothesis. The validity of using the effect size in meta-analysis depends on the validity of the magnitude-insensitivity problem described in Section 5.2. However, it is shown in Section 5.3.2 that the effect size is not related to the degree of evidential support. Moreover, the effect size is irrelevant to theory corroboration when one respects the technical meaning of the term 'effect' (as has been shown in Section 5.7.3). This is the case because the effect of the research manipulation is not the consequence of an efficient cause in the theory-corroboration experiment. In short, given the fact that only qualitative information is required in theory corroboration, the type of quantitative information conveyed by the effect size is irrelevant as an index of the degree of evidential support.

Many more meta-theoretical issues arise if meta-analysis is used as a theory-corroboration tool (e.g., the selection problem, the lack of independence problem, the unjustifiable disregard for the quality of the research, the lack of commensurability among the to-be-aggregated studies, etc.). As these issues are not related to NHSTP, interested readers are referred to Chow (1987b, 1987c), Cook & Leviton (1980), Eysenck (1978), Gallo (1978), Leviton & Cook (1981), Mintz (1983), Rachman & Wilson (1980), Sohn (1980), Wilson & Rachman (1983). Be that as it may, it is necessary to consider the implication of the incommensurability of the to-be-aggregated studies on meta-analysis. It is also instructive to examine the claim that researchers accumulate knowledge, a goal incompatible with the binary nature of NHSTP.

5.8.3. *The Incommensurability Problem in Meta-analysis*

As may be recalled from Section 5.8.1, it is the meta-analytic practice to aggregate data from various studies dealing with the same research subject matter. This meta-analytic practice is wanting in the case of theory-corroboration research for the following reason. Recall from Section 3.2 that the to-be-corroborated explanatory theory implies multiple research hypotheses. Given any research hypothesis derived from the substantive hypothesis, there are many possible experimental hypotheses. Moreover, these diverse experimental hypotheses are about various theoretical properties or functions of the hypothetical mechanism implicated in the to-be-corroborated theory.

The testing of these various experimental hypotheses may implicate different experimental situations. An important feature of any experiment is the choice of the dependent variable, as may be recalled from Section 3.2.3. Important to the present discussion is the fact that different dependent variables may be used in various experiments, even though the experiments are all derived from the same substantive hypothesis. This state of affairs makes it inappropriate to aggregate the effect sizes, or any other quantitative indices, of the various experiments.

Another feature that may render different experiments incommensurable, even though the experiments are about the same theory, is the choice of the independent variable. This may be seen from the entries in the 'Independent variable' and 'Property or function of the iconic store studied' columns in Table 5.5. The entries in these columns show that different independent variables are implicated when different aspects of the same theory are being investigated.

To complicate matters more for meta-analysts, the prescribed $(u_E - u_C)$ may be positive in one case, but negative in another. It may even be zero. For example, Schneider and Shiffrin (1977) used a flat reaction-time function to support their theory of automaticity. This questions the validity of the meta-analytic recommendation that the values of effect size from all studies be aggregated. In short, the combined Z or combined effect size would not be a meaningful indicator of the tenability of the theory. To aggregate data from these studies would be like mixing apples and oranges. Many researchers consider this an illegitimate practice (see, e.g., Cook & Leviton, 1980; Mintz, 1983; Presby, 1978).

Glass et al. (1981) answer this incommensurability criticism by saying that it is proper to mix apples and oranges because they are fruits. That is, incommensurable studies at one level of discourse become commensurable if the studies are subsumed in a higher-order category. However, Glass et al.'s answer is unsatisfactory, because

> [studies] of oranges may have been specifically designed to study unique features of oranges, not because [oranges] are fruits ... For example, the acidity of oranges (or the texture of apples) is not a property common to all fruits. The outcome of Study A may be due to the acidity of oranges; the texture of apples may be the reason for the outcome of Study B. In other words, oranges and apples should not be mixed. (Chow, 1987c, p. 262)

5.8.4.   Knowledge Growth – Evolution or Accumulation?

Corresponding to the data-accumulation procedure described in Section 5.8.1 is the meta-theoretical assumption that the growth of knowledge is a simple matter of information accumulation. Meta-analysts consider the

Table 5.6   *The phenomenon-hypothesis-implication-data (P-H-I-D)*
consistency at different research stages of the substantive hypothesis, H

Experiment	Phenomenon/ prior data	Hypo- thesis	Impli- cation	Data	P-H-I-D consistency
Before experimentation	P	H			Yes
1	P	H	I_1	D_1	Yes
2	$P + D_1$	H	I_2	D_2	Yes
3	$P + D_1 + D_2$	H	I_3	D_3	Yes
...		H	...	...	Yes
...		H	...	...	Yes
$n - 1$	$P + D_1 \ldots + D_{n-2}$	H	I_{n-1}	D_{n-1}	Yes
n	$P + D_1 + \ldots D_{n-1}$	H	I_n	D_n	Yes
...		H	...	...	Yes
t	$P + D_1 + \ldots D_{t-1}$	H	I_t	D_t	Yes

binary nature of NHSTP antithetical to such a process (e.g., Schmidt, 1992, 1994, 1996). However, it can be shown that the growth of knowledge may better be characterized as the evolution of ideas achieved by a trial-and-error process at the conceptual level.

Consider the entries in the 'Effect Size' column in Table 5.5 from the meta-analytic perspective. The conclusion of Section 5.8.3 is that the entries in question come from incommensurable studies. However, meta-analysts simply ignore this incommensurability. To them, the various experimental studies are simply different sources of the same genre of 'raw data'. The data are the same because they are all about the iconic store. In other words, meta-analysts do not consider the following question important:

> **Why are multiple, and qualitatively different, studies undertaken to corroborate the same substantive hypothesis?**
>
> [Q5-4]

However, Question [Q5-4] is very important from the theory-corroborative perspective for the following reason. It may be recalled from the 'Property or function of the iconic store studied' columns in Table 5.5 that various properties or functions of the iconic store are being investigated with the series of 12 experiments. That is to say, instead of investigating the hypothetical mechanism *iconic store* directly, experimental psychologists test theoretical statements about various qualitatively different observable consequences that should be brought about by the theoretical properties of the unobservable iconic store.

The rationale underlying the 12 studies summarized in Table 5.5 may be illustrated with reference to Table 5.6. For ease of exposition, 'Hypothesis H' and 'Phenomenon P' in this table refer to the iconic-store hypothesis and the phenomenon of perceiving more than can be recalled (Sperling, 1960), respectively. Further, suppose that Implications $I_1, I_2, \ldots, I_t$ in the table represent, respectively, theoretical statements about 'Rate of decay',

'Independence of location and identity information', . . . , 'No identity information' in Table 5.5. For example, Implication I_1 refers to the theoretical expectation that information in the iconic store decays systematically as the inter-stimulus interval (ISI) increases.

As may be seen from the 'Before experimentation' row of Table 5.6, the to-be-explained phenomenon exists before any attempt to theorize about it. The explanatory theory H (see the 'Hypothesis' column) is proposed to account for the phenomenon. Hence, the necessary requirement is that H be consistent with P. However, *vis-à-vis* P, H is an *ad hoc* hypothesis, for the simple reason that H is formulated for the specific purpose of explaining P. Consequently, the truth of H is not enhanced by accumulating more occurrences of P.

In its capacity as an explanatory hypothesis, what is said in Hypothesis H must be more than a description of Phenomenon P. In other words, hypothetical structures with specific theoretical properties or functions must be explicitly stated in H. For example, two generally accepted properties of the iconic store are (a) its relatively large capacity (i.e., larger than that of the short-term store), and (b) the rapid decay of iconic information (viz., within 500 msec). Each of these theoretical properties has to be corroborated (viz., Experiments 1 and 2 in Table 5.5 for the decay and capacity assumptions, respectively). Moreover, it must be possible, with Hypothesis H, to account for phenomena other than P if Hypothesis H is to be more than a description of P (e.g., Experiments 5 and 12 in Table 5.5).

Consider Row 1 in Table 5.6. It depicts that the first attempt to test the substantive hypothesis, H, is to work out the first implication of H, namely, I_1. Implication I_1 must be consistent with P; but it need not be consistent with the rest of the belief system.[6] Testing of this implication is done by setting the data-collection condition such that what is said in I_1 should[7] occur if H is true. Hence, research data (D_1) are matched against I_1. The hypothesis is considered tenable if there is a match (see Section 4.4).

Another implication, I_2, of H is worked out at the onset of Experiment 2. The necessary requirement before conducting the research is that I_2 be consistent with the conjunction of P and D_1. By the time the research reaches Experiment n, Implication I_n has to be consistent with the collectives made up of P, D_1, D_2, . . . D_{n-1}. The important point to note is that Hypothesis H is not tenable if any one of its implications is not substantiated by data. This is contrary to the assertion that it is incorrect to entertain 'the assumption that each individual study must be able to support and justify a conclusion' (Schmidt, 1994, p. 19).

In other words, it is emphasized in the theory-corroboration approach that the various data sets (viz., D_1, D_2, . . ., D_{n-1}) depicted in Table 5.6 may be, and often are, qualitatively different kinds of observation (see Table 5.5). Each set of data is compared with a particular implication of Hypothesis H. Moreover, the series of experiments represents different attempts to falsify Hypothesis H from various angles. To the extent that they fail to refute the theory, they converge on the tenability of the theory. Hence, the set of experiments constitutes collectively the converging operations for establishing the tenability of the theory (Garner, Hake & Eriksen, 1956).

As the series of converging operations progresses, our understanding of Phenomenon P increases. In this sense, the advance of knowledge assumes the form of refining, with valid evidential support in each step, what is said in the explanatory theory. This process of refinement is not the accumulation of quantitative information (and definitely not the accumulation of the same type of data) because qualitatively different data from various experiments are kept separate. Instead, it assumes the form of (a) eliminating, or modifying, incorrect theoretical statements, and (b) refining correct theoretical statements by explicating the theoretical properties of the hypothetical mechanisms.

In sum, the growth of knowledge is a process of trial and error at the conceptual level. Hence, it seems to be more like a deliberate evolution of empirically substantiated ideas than a mechanical accumulation of data. That is,

> [Explanatory theories] are not cumulative or generalized descriptions of facts, observations, or empirical laws, and cannot be discovered, developed, or evaluated by concatenating descriptions of different experimental paradigms. Although [explanatory theories] summarize a wide range of empirical generalizations, they do not directly describe events specific to particular experimental paradigms or situations. (MacKay, 1993, p. 249)
>
> [P5-1]

The present discussion is restricted primarily to the role of p and of the effect size in the theory-corroboration procedure envisaged in meta-analysis. The difficulty with the meta-analytic treatment of p is that the conditional nature of p is not given its proper recognition in meta-analysis (see Section 2.8.2). The meta-analytic assumption about the contribution of the effect size to theory corroboration is also debatable. An obvious difficulty is the failure in the meta-analytic approach to distinguish between theory corroboration and statistical decision making.

5.9. Summary and Conclusions

The *ambiguity–anomaly* criticism of NHSTP is made up of the sample size-dependence, the incommensurate significance-size, and the magnitude-insensitivity problems. These issues seem to have more to do with the putative cynicism or cavalier attitude that may be assumed by researchers than with the intrinsic difficulties with NHSTP as a research tool. What critics have ignored is the fact that researchers are constrained by the need to achieve internal validity and external validity in their studies. Moreover, in choosing the α level by convention, NHSTP users also commit themselves to other concomitant methodological obligations.

There are reservations about using the effect size as the numerical index of the real-life importance of the research result. The chance-versus-size problem is the result of the conflict between the connotative and technical meanings of the terms 'effect' or 'effective'. A case is made that there is no justification for adopting the connotative meaning of 'effect' or 'effective' when assessing data from the theory-corroboration experiment. In the case of the utilitarian experiment, appealing to the connotative meaning of 'effect' actually introduces extra-statistical considerations.

Underlying the proposal to assess the result of empirical research in terms of the effect size are two assessment criteria, namely, the statistical criterion for assessing chance influences, and some non-statistical criteria for evaluating the real-life impact of the result. It becomes necessary to have an account of the interplay between the two criteria. However, this account is not available.

NHSTP users would not object to applying non-statistical criteria after chance influences are ruled out by virtue of the fact that statistical significance is obtained. However, effect size advocates are prepared to apply the non-statistical criterion even when it is not possible to exclude chance influences as an explanation of the result. Why is statistics used at all? Should the researcher embark on a course of action when the research outcome may actually be the result of chance influences?

The earlier explication in Chapter 2 of the conditional nature of p and the qualitative criterion of theory corroboration in Section 5.3 suggest some reservations about meta-analysis as a theory-corroboration procedure. For example, meta-analysts assume that p is informative as to the tenability of the substantive hypothesis and that the size of the effect is somehow indicative of the evidential support for the theory. Neither of these two assumptions is correct.

Notes

1. This is an index of the effect size suggested by J. Cohen (1987). The effect size is sometimes indicated by the proportion of variance accounted for (r^2) or the Pearson correlation coefficient between the independent and dependent variables, r (see, e.g., Rosnow & Rosenthal, 1989). None the less, the spirit of the present discussion is applicable to any index of the effect size.

2. The essence of the partial-report task is to overload the subject's visual system during the extremely brief stimulus presentation. However, the subject is told to recall only a portion of the information at various delays after the offset of the stimulus. A typical partial-report trial is made up of the following sequence of events. Nine letters are presented in three rows of three for 50 msec. At various delays after the nine-item display is withdrawn, a tone is presented. The delay in question is the 'inter-stimulus interval' or ISI. The subject has to recall only the top (high tone), middle (medium tone) or bottom row (low tone) of letters. The points to emphasize are that the subject does not know which row of letters to recall when the stimulus is still in sight, and that the recall requirement is within the subject's memory span.

3. The obligation imposed on NHSTP users by the conventional nature of the α level seems unimportant to some critics. For example, it is said, 'Unfortunately, [the status of the α level] as only a convention is frequently ignored; there are many published instances where a researcher, in an effort at rectitude, fails to report that a much desired null rejection would be possible at the .06 level but instead treats the problem no differently than he would have had it been at the .50

level' (J. Cohen, 1987, p. 12). The undertone of this comment is disturbing. Why is rectitude objectionable when the decision criterion is well-defined and objective? For what reason should the researcher's desire be taken into account when the issue is about methodological rigour? Why is the said desire germane to the validity of the argument about the evidential support for the hypothesis afforded by the research data (see Section 5.3.3)?

4. Strictly speaking, to say this is to concede too much. Consider $t = [(\bar{X}_1 - \bar{X}_2) - (u_1 - u_2)]/s_{(\bar{x}_1 - \bar{x}_2)}$. It means that the difference between the two means is expressed in standard error units. The difference between t and J. Cohen's (1987) d is that, whereas the denominator of the index of the effect size (d) is the standard deviation of either the experimental or control population, the denominator of the t statistic is the standard error of the difference. This discrepancy may be the reason why the apparent anomaly between statistical significance and the effect size arises in the first place.

5. The distinctions among the efficient, material and formal causes discussed in this section are three of the four kinds of cause identified by Aristotle. See Loomis (1971).

6. It is the Popperian view that theoretical conjectures are better the bolder they are. If researchers restrict their ideas only to those consistent with existing beliefs, there will be no room for new theoretical insights. However, there are two important codas to the Popperian view. First, the onus is on the researcher to substantiate the bold conjectures by subjecting them to rigorous falsification attempts. It is for this reason that methodological rigour is important. Second, having succeeded in upsetting the established system of beliefs, the part of the belief system that is affected should be revised. This revision of the belief system is then subjected to a rigorous falsification process.

7. See n. 2 in Chapter 3 for the meaning of 'should'.

6

A Critical Look at Statistical Power

Power analysis is debatable because (i) the probability, $(1 - \beta)$, is an unknown value, (ii) the conditional nature of statistical power is not acknowledged in assertions about statistical power, (iii) H_1 is treated incorrectly as the substantive hypothesis, and (iv) the putative relationship between statistical power and statistical significance is more apparent than real. Additional meta-theoretical difficulties with power analysis are seen by examining the similarities and differences between NHSTP and the theory of signal detection (TSD). Statistical power may say something about the researcher, but not about the substantive hypothesis. Also questioned are (a) the validity of assigning a probability to statistical significance, and (b) the appropriateness of treating questions about the tenability of the substantive hypothesis as a statistical concern. It is also questionable to incorporate utilitarian decisions into a statistical procedure.

6.1. Introduction

To power analysts, the *ambiguity–anomaly* problem discussed in Section 5.2 would not occur if the power of the test were known. Specifically, the researcher can be assured of the validity of a significant result if the test is of sufficient power. On the other hand, an insignificant result can be safely ignored if the test is of sufficient power. Specifically, it is said,

> from a power analysis at, say, $\alpha = .05$, with power set at, say, .95, so that $\beta = .05$, also, the sample size necessary to detect this negligible effect with .95 probability can be determined. Now if the research is carried out using that sample size, and the result is *not* significant, as **there had been a .95 chance of detecting this negligible effect**, and the effect was *not* detected, the conclusion is justified that no nontrivial effect exists, at the $\beta = .05$ level. (J. Cohen, 1990, p. 1309; emphasis in boldface added)

[Quote 6-1]

Hence, power analysts encourage empirical researchers (particularly psychologists) to take into account statistical power when they choose the sample size. Researchers should also supplement their report of statistical significance with an explicit specification of the power of the test carried out.

There is another reason why statistical power is said to be important. It is a power analytic thesis that the proof a theory or a substantive hypothesis is established if the effect implied by the theory is detected (see [Quote 6-1] above). At the same time, the effect is detected if the result is statistically significant. The probability of obtaining statistical significance is determined by the power of the statistical test, as may be seen from the following assertion:

> The power of a statistical test is the probability that it will *yield* statistically significant results. (J. Cohen, 1987, p. 1; emphasis added)

[Quote 6-2]

There are two parts to the present discussion of power analysis. It is argued in the first part that power analysis is problematic even if the validity of the concept statistical power is not questioned. The validity of statistical power itself is questioned in the second part. It is pointed out that NHSTP users and critics define 'Type II error' differently. The mechanical component of power analysis is then examined in terms of the power analytic solution of the *ambiguity–anomaly* criticism of NHSTP. It is shown that statistical power and statistical significance cannot be related in the way envisaged in the power analytic account. This assessment is achieved by making explicit power analysts' implicit appeal to the theory of signal detection (TSD). The conceptual difficulties with the concept statistical power are shown in the second part by exploring the implications of some assertions about statistical power.

6.2.   The Ambiguity of Statistical Significance – Mathematical or Methodological?

Power analysts accept the *ambiguity–anomaly* criticism of NHSTP discussed in Section 5.2. That is, the result of a test of significance without power specification is ambiguous to power analysts because an insignificant result may be due to insufficient power as a result of the fact that the sample is too small. On the other hand, the result may be statistically significant simply because too large a sample is used. The power analytic suggestion is that an appropriate sample size can be determined by consulting the Sample Size Tables (J. Cohen, 1987). An alternative set of tables may be found in Kraemer and Thiemann (1987).

Consider the assertions that statistical power is increased when the sample size is increased, and that increasing the power of a test increases the chance of obtaining statistical significance. The standard error of the difference becomes smaller when a larger sample size is used because the denominator of the equation of the standard error is the square root of the sample size (one-sample case) or of some function of the sizes of the samples involved (two-sample case). The test statistic consequently becomes larger with increases in sample size. In other words, assuming that the numerator remains constant, it is easier for a test statistic based on a larger sample to reach a pre-determined criterion of statistical significance than it is for one based on a smaller sample. However, this state of affairs is brought about by the equation. It has nothing to do with statistical power because the standard error is reduced when sample size is increased whether or not statistical power is evoked.

At the methodological level, issues about the sample size arise because of questions about data stability or reliability. At the same time, these questions are methodological or design issues. It may be suggested that the true solution of the ambiguity of statistical significance or non-significance lies not on another numerical index, but on some safeguards in the data collection procedure. The real issue is about the inductive conclusion validity of empirical research (see Section 4.7). It also helps to note that questions about the relationship between the inductive conclusion validity and the statistical conclusion validity are not numerical issues, a point also made by Neyman and Pearson (1928) as follows:

> The sum total of the reasons which will weigh with the investigator in accepting or rejecting the [substantive] hypothesis can very **rarely** be expressed in numerical terms. All that is possible for him is to balance the results of a mathematical summary, formed upon certain assumptions, against other less precise impressions based upon *à priori* or *à posteriori* considerations. (Neyman & Pearson, 1928, p. 176; emphasis in boldface added)
>
> [Quote 6-3]

An example of Neyman and Pearson's (1928) 'numerical terms' may be $(1 - \beta)$ namely, power. An example of the à priori consideration implicated is experimental psychologists' careful choice between using a between-groups or a repeated-measures design. In addition to the sample size, experimental psychologists (particularly cognitive psychologists) also pay attention to the number of trials given to each subject in every treatment combination. There is also the question as to the amount of training on the experimental task provided for the subjects before data collection. These deliberations are undertaken with reference to theoretical and method-ological considerations (see Sections 5.2.2 and 5.2.3; see also Frick, 1995).

An example of Neyman and Pearson's (1928) à posteriori consideration may be illustrated with reference to the schematic representation of method of difference in Table 4.4. Recall that it is assumed in the experimental analogue of method of difference that the extraneous variables E1–En are held constant at both levels of the independent variable by using the same subject in both conditions (see Section 4.5.1). An alternative means to exclude extraneous variables is to assign randomly to the two conditions subjects who are comparable in terms of a set of theoretically relevant criteria.

However, it is possible that an extraneous variable is discovered at the conclusion of the study to have varied systematically with the two levels of the independent variable. The extraneous variable so identified becomes a confounding variable of the study. The internal validity conclusion of the study is thereby compromised by the confounding variable. The attempt to ensure that there is no confounding variable in the completed research is an example of Neyman and Pearson's (1928) à posteriori consideration.

The important point is that both the à priori and à posteriori considerations are concerns about the data-collection procedure or the logic of the experimental design. There is nothing 'less precise' about these considerations, despite Neyman & Pearsons' (1928) characterization. They are not reservations about the numerical value of a statistical index, be it α, p, the calculated t or $(1 - \beta)$. In short, the ambiguity of statistical significance or non-significance is a methodological, not a statistical, issue. (See the discussion of Swain et al.'s, 1990, oat-bran study in Section 4.8 for an illustration.) The ambiguity cannot be resolved with a numerical index such as statistical power.[1]

6.3.   The Mechanical Nature of Power Analysis

Determining the sample size is a mechanical exercise in power analysis because a researcher needs only three numerical parameters to use the *Sample Size Tables*: (a) the α level required, (b) the effect size to be detected, and (c) the statistical power desired (J. Cohen, 1987, 1992b). Although 'mechanical' is not a derogatory characterization, the mechanical nature of power analysis is examined here because the *Sample Size Tables* are general-purpose tables to be used regardless of research type (viz., correlational, quasi-experimental or experimental), research objective (i.e., utilitarian or theory-corroborative) or methodological concerns (e.g., the nature of experimental task, independent variable, dependent variable). Is the determination of sample size really as mechanical as envisaged in power analysis? Can the *ambiguity–anomaly* criticism of NHSTP be satisfactorily dealt with in a mechanical manner?

A function of increasing the sample size is to stabilize the data. Data stability is indicated by a small standard error of the test statistic. However, increasing the sample size is not the only means of stabilizing the standard error. Nor is it the proper one to use under some circumstances. For example, small sample sizes are often used in studies of the iconic store (e.g., Sperling, 1960; Section 5.2.3 above). The data are stabilized by giving the few subjects extensive training on the partial-report task before data collection (see n. 2 in Chapter 5; Chow, 1985). That is, the tradeoff between using fewer well-trained subjects and using a larger number of inexperienced subjects is not recognized in the power analytic dependence on the *Sample Size Tables*.

It also seems that the distinction between the between-groups and repeated-measures design is not recognized in power analysis. In general, fewer subjects are required for the repeated-measures design than its completely randomized counterpart when it is appropriate to use the repeated-measures design (e.g., when the order of testing effect can be minimized). In short, some methodological considerations in empirical research render questionable the mechanical way of determining sample size in power analysis.

In subscribing to the *ambiguity–anomaly* criticism of NHSTP, power analysts are also concerned with the researcher's probable cynicism or cavalier attitude discussed (see Section 5.2.1). The crucial question is whether or not a mechanical, numerical exercise (such as power analysis) can prevent researchers from falling into either scenario. This is particularly serious when it is easy to misuse the *Sample Size Tables* because the choice of the to-be-detected effect size or the decision about the statistical power required is arbitrary (J. Cohen, 1987).

Consider the case of desired power, the choice of which is guided by substantive concerns. However, these substantive concerns are not necessarily methodological or theoretical considerations because it is said that, '[by] studying the relationship between *n* and power for his situation, taking into account the increase in cost to achieve a given increase in power, he can arrive at a rational solution to the sample-size problem' (J. Cohen, 1965, p. 98). It is not clear how this guideline would prevent someone from adopting a cynical or cavalier attitude when there is no independent, well-defined criterion for assessing the cost or rationality. Moreover, cost (however it is defined) has nothing to do with whether or not a research conclusion is warranted by the data collected for the specific purpose of testing the substantive hypothesis.

Recall that the standard error term used in tests of significance (e.g., the standard error of the difference) is sensitive to, among other things, the research design, the nature of experimental task, the choice of independent and dependent variables (as well as the specific levels of the independent variable), and the amount of practice available to subjects. How is one to determine the appropriate level of statistical power or effect size for a given study under such circumstances? Furthermore, important differences among correlational, quasi-experimental and experimental studies render it impossible to rely on general-purpose indices (including statistical power) to settle non-statistical issues (e.g., the inductive conclusion validity of the study). What situational factors are important determinants of desired power? In short, the mechanical aspect of power analysis is not suited for preventing cynicism or the cavalier attitude. Nor can using power analysis ensure that the formal requirement of Mill's (1973) inductive methods is satisfied (see Section 4.5). In short, the usefulness of the general-purpose *Sample Size Tables* is debatable as a tool to ensure that non-statistical concerns are met.

6.4. Power Analysis and the Theory of Signal Detection (TSD)

As may be recalled from any panel in Figure 2.4, the rationale of NHSTP is fully depicted with one sampling distribution (see, e.g., Hopkins et al., 1996). However, if it is also necessary to represent statistical power, the graphical representation of NHSTP requires two distributions. This may be seen from the top panel of Figure 6.1. The present juxtaposition the top and bottom panels of Figure 6.1 makes it easier to discuss an implicit appeal to TSD made by power analysts (e.g., J. Cohen, 1965, 1987; Edwards, Lindman, & Savage, 1963). Some difficulties with power analysis may also be seen more readily when the putative affinity between NHSTP and TSD is made explicit and examined. An attempt is made to show that these problems make it doubtful that power analysis is as useful as envisaged by power analysts.

6.4.1.   *Truth and Detectability in Power Analysis*

To begin with, the statistical alternative hypothesis, H_1, is identified with the hypothesis about Phenomenon P in power analysis. That it is misleading to do so has been shown in Section 3.2 above. None the less, establishing the tenability of a theory is treated in power analysis as no different from detecting a signal in the psychophysics situation. For example, it is asserted that, if Phenomenon P exists, its effects must be detectable (see, e.g., J. Cohen, 1987). That is to say, the evidence for H_1 is the detectability of the aphenomenon itself. At the same time, to detect a difference is to obtain statistical significance. Hence, statistical significance is indicative of the truth of H_1 or the fact that Phenomenon P exists. For this reason, it is important to power analysts to know the a priori probability of obtaining statistical significance (J. Cohen, 1987). The said a priori probability is supplied by the power of the test (viz., [Quote 6-2]; see also Mosteller & Bush, 1954).

This is the reason why the power of the statistical test is treated as the probability of H_1 being true (by virtue of the fact that it represents, to power analysts, the probability of obtaining statistical significance). Researchers are urged to use as powerful a statistical test as possible because a more powerful test affords a higher probability of obtaining statistical significance. Power analysis is, however, debatable at the conceptual level. Some difficulties may be shown by contrasting it to TSD.

6.4.2.   *The TSD Overtones in Power Analysis*

The rationale of inferential statistics is described in some versions of power analysis in a way no different from that of TSD. For example, it is said that 'effects are appraised against a background of random variation' (J. Cohen, 1987, p. 13), and that the said appraisal consists of '(*detecting*) a difference between the means of populations A and B . . .' (J. Cohen, 1987, p. 6; emphasis added). As another example, operating characteristics refers to an index of an observer's performance in a TSD task situation (viz., receiver operating characteristics or ROC; Green & Swets, 1966; McNicol, 1972). In the context of NHSTP, operating characteristics is an a priori means of

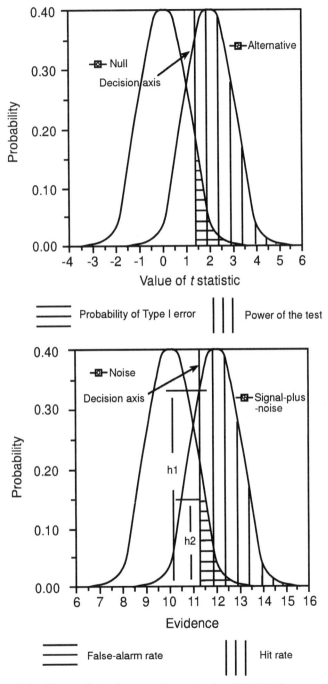

Figure 6.1   *The similarity between the rationale of NHSTP (top panel) and the theory of signal detection (TSD) (bottom panel)*

representing a researcher's decision making in power analysis (Ferris, Grubbs, & Weaver, 1946; Mosteller & Bush, 1954).

As for the rationale of power analysis, suggestive of the implicit affinity between NHSTP and TSD is the appeal made in power analysis to Neyman & Pearson's (1928) sample-to-population approach. That is, researchers first determine the sample statistic, $\overline{X}$, and then ask what the probability is that the sample is selected from Population P with parameter u. An appeal to the a posteriori probability is found in a TSD analysis. Similarly, an important role for a posteriori probability is assumed in power analysis, as witnessed by the following assertion:

> Now, what really is at issue, what is always the real issue, is the probability that H_0 is true, given the data, $P(H_0|D)$, the inverse probability. (J. Cohen, 1994, p. 998)
> [Quote 6-4]

The brightness detection task in the psychophysics situation may be used to show some superficial similarities between NHSTP and TSD. An observer fixates at a point on the lit screen. On hearing a tone, the observer decides whether or not an additional light has been presented at the fixation point. The additional light, if it is presented, varies in intensity from trial to trial. The two mutually exclusive and exhaustive states of affairs are (a) an additional light is presented (i.e., the signal event[2]), and (b) no additional light is presented (i.e., the noise event). The stimulus intensity at the fixation point accompanying the probe tone on a trial is the to-be-judged evidence.

The observer's task is to decide whether a signal or noise event is indicated by the evidence. A hit is scored when the observer responds 'Yes' to a signal event. The observer commits a false alarm when a 'Yes' response is made to a noise event. The prior odds (i.e., the a priori ratio of the frequency of noise events to the frequency of signal events) is represented by two overlapping normal distributions with the same variance, as depicted in the bottom panel of Figure 6.1. Represented on the horizontal-axis are the possible values of the evidence (the stimulus intensity).

Suppose that the to-be-studied phenomenon is P. Analogous to the noise and signal events in the TSD situation are P does not exist and P exists, respectively, in power analysis.[3] These two mutually exclusive and exhaustive states of affairs are characterized as true H_0 and true H_1, respectively. The test statistic obtained at the end of data analysis would be the evidence, and a researcher's task is to decide whether the evidence is indicative of the statistical alternative hypothesis (H_1), or the null hypothesis (H_0). The distributions underlying H_0 and H_1 are also represented by two overlapping normal distributions, as shown in the top panel of Figure 6.1.

6.5. Issues Raised by the NHSTP–TSD Affinity

A correspondence between two sets of descriptive terms becomes obvious if the affinity between NHSTP and TSD is assumed. Of particular interest is the affinity between statistical power and hit rate, as may be seen from the areas shaded with vertical lines in both panels of Figure 6.1. Making this correspondence explicit renders it easier to see what is questionable with [Quote 6-2] in Section 6.1 above.

6.5.1. Statistical Power – A Conditional Probability

In addition to what is said in Section 2.2.4, there is another way to show that statistical power is a conditional probability. A Type I error is made when a researcher rejects H_0, given that H_0 is true. This is analogous to committing a false alarm in TSD. The areas shaded with horizontal lines in the top and bottom panels of Figure 6.1 represent the probability of Type I error in NHSTP and false-alarm rate in TSD, respectively. The complement of Type II error (i.e., the correct response of rejecting H_0, given that H_1 is indeed true) in power analysis is the equivalent of a hit in TSD. At the same time, hit in TSD refers to a 'Yes' response *contingent on* the presence of a signal event. Hence, *hit rate*, is a conditional probability. As a conditional probability, it says nothing about the exact probability of the presence of a signal event.

Analogous to the presence of a signal event in TSD is the presence of a true H_1 in power analysis. As the probability of a signal event is an exact probability, the probability of H_1 being true should also be an exact probability. However, the TSD analogue of statistical power is the conditional probability, hit rate. Hence, if it were meaningful to talk about the probability of accepting a true H_1 (viz., statistical power), it would be meaningful only as a conditional probability (Chow, 1991c). The conditional nature of probability statements in NHSTP may be seen from the following assertion:

> the immediate statistical outcome of any significance test can be thought of as a *conditional probability*, the probability of a result this much or more deviant from expectation in a particular way **given** that the hypothesis tested is really true'. (Hays, 1963, p. 247; emphasis in boldface added)
>
> [Quote 6-5]

Consequently, the power of a test (a conditional probability) does not inform us as to the probability that a theory is true (an exact probability), even if the truth of the substantive hypothesis were a matter of probability or a matter of signal detection. This is analogous to not knowing the probability of the signal event (an exact probability) even though the hit rate (a conditional probability) is known. Hence, what is said in [Quote 6-2] is questionable.

Furthermore, empirical research is conducted to corroborate theories in the sense of assessing whether or not there is evidential support for the to-be-tested theory. This 'warranted assertibility' (Manicas & Secord, 1983, p. 410) is an all-or-none issue, for which a statement in probabilistic terms (be it exact or conditional) is inappropriate (Chow, 1989). In sum, theory corroboration is not a detection task. Important to the present discussion is that statistical power is not informative as to whether or not a theory is warranted by data.

6.5.2. *An Index of the Truth of H_1 or the Researcher's Confidence?*

An observer's hit rate, as a conditional probability, tells us something about the observer (viz., the strength of the signal as it appears to the observer), not the strength of the signal as a physical attribute. Similarly, in view of Neyman and Pearson's (1928) sample-to-population approach to inferential statistics, statistical power may be treated as an index of the a posteriori confidence of the researcher about the statistical decision,[4] but not as an index of the a priori probability that H_1 is true. This is contrary to what is said in [Quote 6-2] or the part in [Quote 6-1] highlighted in boldface. While there is nothing wrong in making explicit a researcher's confidence, this information is not indicative of the extent to which the theory is warranted by data.

6.5.3. *Statistical Power – Cumulative Probability or Efficacy?*

[Quote 6-1] suggests that statistical significance is reached by virtue of the numerical index statistical power. (See also the emphasis in italics in [Quote 6-2].) It suggests that a particular efficacy or capability is made available to the statistical procedure (viz., the efficacious agency responsible for yielding statistical significance) if the sample size is determined with reference to the appropriate *Sample Size Tables* (J. Cohen, 1987). This assertion of efficacy is misleading because, in terms of the graphical representation shown in the top panel of Figure 6.1, statistical power refers to the cumulative probability over a range of values of the test statistic, *contingent on* the fact that chance influences are rejected (viz., the area of the 'Alternative' distribution to the right of the decision axis shaded vertically in the top panel of Figure 6.1).[5] That is, no efficacy or capability of any sort is implicated at this level of discourse. A conceptual justification is required if a putative efficacious agency is attributed to the numerical index statistical power. As no such justification is offered, it seems more appropriate not to attach any extra-statistical meaning to the term statistical power.

There is no a priori reason why the decision to reject the false H_0 should not simply be called Type II correct decision. Power analysis might not be so readily accepted had the vertically shaded area in the top panel of Figure 6.1 been labelled with a non-evocative term like not-β or probability of Type II correct decision. In other words, an unwarranted meaning is attributed to statistical power, which is a conditional probability, as a result of its being given the evocative label. A connotative meaning of 'power' is efficacy or capability. The same, of course, may be said of *significance*, a connotative meaning of which is importance or impact. It is possible that many of the debatable things said about statistical significance had been said because of the unwarranted connotative meanings of the word 'significance'. By the same token, many of the criticisms of NHSTP might not have been raised had 'not chance' and 'chance' been used for 'significance' and 'not significance', respectively (see Section 3.3).

6.6. Dissimilarities between NHSTP and TSD

Despite the apparent similarities between NHSTP and TSD, these two procedures are different in many respects. The validity of the TSD technique depends on satisfying a set of methodological and procedural requirements. Hence, additional difficulties with power analysis become obvious if it is realized that many of these TSD requirements are not met when testing a statistical hypothesis.

6.6.1. Evidence – Well-defined versus Ambiguous

As may be recalled from Section 6.4.2, one of two unambiguous states of affairs is presented to the observer in a TSD task (i.e., a signal or a noise event). The identity of the evidence is unambiguous because it can be assessed objectively with reference to the actual stimulus used on a trial. Consequently, it is possible for the researcher to characterize unambiguously the observer's responses as hits, correct rejections, false alarms or misses. Moreover, the evidence-assessment criterion in TSD is independent of the TSD procedure itself.

The evidence in NHSTP comes in the form of a relation between two or more sample statistics at the conclusion of the research (e.g., $\overline{X}_1 - \overline{X}_2$). There is no objective basis to decide whether the difference obtained is indicative of 'P exists' or 'P does not exist'. As a matter of fact, research is conducted precisely for the reason that the identity of the evidence (viz., a chance difference or otherwise) itself is uncertain. NHSTP itself is the evidence-assessment procedure. In short, the statistical decision confronting a researcher in a NHSTP situation is very different from that presented to an observer on a TSD trial.

6.6.2. Multi- versus Single-trial

An observer's sensitivity and response bias are calculated on the basis of a large number of observations in the TSD task situation. Suppose that a rating-scale task is used (i.e., the observer is asked to indicate, in addition to responding 'Yes' or 'No', how confident the observer is in making the response). It is suggested that at least 250 trials of the signal and 250 trials of the noise events are required to determine the parametric index of sensitivity, d' (McNicol, 1972).

Using a one-factor, two-level design, the researcher obtains a difference between two sample means at the end of the statistical analysis. This effectively means that the researcher is being asked to make a signal-detection decision on the basis of the evidence from a single trial if NHSTP is treated as a TSD task. This shows that TSD concepts do not have valid analogues in the NHSTP context, apparent similarities between TSD and NHSTP notwithstanding.

6.6.3. *Payoff Matrix*: *Measurement versus Theory-Corroboration*

An important manipulation in the TSD task situation is the payoff matrix for an observer, namely, the gains of making a correct response (viz., a hit or a correct rejection) and the costs of committing an error (viz., a false alarm or a miss). If the observer is allowed only two responses (viz., 'Yes' or 'No'), a large number of trials is required at each of several levels of the payoff matrix. This manipulation is necessary for determining the observer's response bias (viz., the observer's willingness to respond 'Yes' when uncertain).

The observer's response bias in TSD is represented by the location of the decision axis vis-à-vis the mean of the noise distribution. This location is determined after data collection by the ratio of the height of the decision axis vis-à-vis the a posteriori 'signal' distribution (viz., h1 in the bottom panel of Figure 6.1) to the height of the decision axis vis-à-vis the a posteriori 'noise' distribution (i.e., h2 in the bottom panel of Figure 6.1). This a posteriori ratio is compared to either one of two a priori ratios.

First, if the observer sets the task objective to maximize the number of correct responses in a large number of test trials, the a priori ratio is simply the prior odds (viz., the ratio of the probability of noise events to the probability of signal events). This task objective is not applicable to a researcher using NHSTP, simply because the probability of Phenomenon P's existence (or the substantive hypothesis being true) is not known.

Second, a different a priori ratio is involved if the objective of an observer in a TSD task is set to maximize gains. The location of the decision axis is determined by the product of the prior odds multiplied by the payoff matrix (see McNicol, 1972). In other words if NHSTP were a TSD task, a researcher using NHSTP would attach values to (a) perceived gains of the correct rejection of H_0 or correct acceptance of H_1, and (b) perceived costs of the incorrect rejection of H_0 or incorrect acceptance of H_1.

Indicative of this implicit assumption is the suggestion in power analysis that the placement of the decision axis should reflect a balance struck between statistical power and the α level (J. Cohen, 1987, p. 5). It is said that this may be achieved by taking into account the ratio of the probability of Type II error to the probability of Type I error. This ratio is the NHSTP analogue of response bias in TSD. Researchers are further urged to pay attention to 'the relationship between n and power *for* [*their*] *situation*, taking into account the increase in cost to achieve a given increase in power . . .' (J. Cohen, 1965, p. 98). In sum, flexible placement of the decision axis prior to, or after, data collection is deemed acceptable in power analysis.

The TSD task is a measurement exercise from the researcher's perspective (viz., to measure an observer's sensitivity and response bias; see Macmillan, 1993). That an observer may deviate from the point of no bias is part and parcel of what is to be measured. This is the case because an observer's response bias is compared with the impartial decision of an ideal observer. A lack of objectivity on the part of an observer does not affect the objectivity of the researcher who measures the observer's sensitivity and response bias.

The situation is very different in the case of NHSTP. To begin with, there is no objective payoff matrix in the case of NHSTP (if we ignore, for the moment, the difficulty brought about by the fact that there is no NHSTP equivalent of the prior odds of TSD). The payoff matrix of an individual researcher might be highly idiosyncratic. Moreover, this payoff matrix is not related to the theory–data relationship. This state of affairs is antithetical to objectivity for the following reason.

On the one hand, as a statistical decision maker, the researcher has to be objective because the decision provides the minor premiss of a chain of syllogistic argument, the purpose of which is to assess whether or not a theory is warranted by data (see Section 4.3). The guiding criteria should be only those concerned with the internal and external validity of an empirical study (Campbell & Stanley, 1966; Cook & Campbell, 1979). On the other hand, to take into account a payoff matrix in placing the decision axis is to introduce extra-statistical considerations into statistical decision making. This introduces effectively non-conceptual or non-methodological considerations into the theory-corroboration process. It is a serious threat to objectivity because extra-conceptual concerns have nothing to do with conceptual rigour (see Chapter 5; Chow, 1991a).

To recapitulate, the power analytic thesis is that whether or not the substantive hypothesis is true (as represented by H_1) is established when its predicted effect is detected. The detection of the said effect is indicated by statistical significance. The a priori probability of obtaining statistical significance is said to be given by the power of the test. This power analytic thesis is questioned because H_1 does not represent the substantive hypothesis. Furthermore, as statistical power is a conditional probability, it cannot inform the exact probability of any event, including the probability of obtaining statistical significance.

6.7.   Type II Error and Power

The validity of statistical power is taken for granted in the foregoing discussion. However, there are reasons to question the validity of statistical power itself. First, the meaning of 'Type II Error' is changed when statistical power is introduced. Second, it is not possible to represent statistical power graphically without misrepresenting NHSTP.

First, consider the meanings of Type II Error with and without accommo-dating statistical power. Recall from Section 2.2.4 that Type II error is accepting H_0, given that H_0 is false (see Table 2.2). That is, a Type II error is a conditional event; it is committed when the researcher explains the result with chance influences *in the event that* the hypothesis of chance influences is incorrect. More importantly, only one sampling distribution is implicated in NHSTP, namely, the one predicted on H_0 (see Sections 2.7.1 and 3.2.5). The probability of a Type II error (i.e., β) is a conditional probability, and it should be defined as '$p(\text{Accept Chance}|\text{not-}H_0)$'.

However, the probability of a Type II error is defined in power analysis as '$p(\text{Accept Chance}|H_1)$', as may be recalled from Figure 2.1. This power analytic definition of 'Type II error' may be called 'Type II error$_p$'. This is problematic because it is incorrect to treat H_1 as synonymous with 'not-H_0', although H_1 and H_0 are mutually exclusive and exhaustive alternatives (see the lower panel of Table 2.1). Seen in this light, '$p(\text{Accept Chance}|H_1)$' and '$p(\text{Accept Chance}|\text{not-}H_0)$' are conditional probability statements about different events. The former is more specific than the latter. At the same time, there is no justification for the more specific statement.

The point to emphasize is that while H_1 plays an important role in power analysts' definition of β, it plays no role in the definition of 'Type II error' adopted in NHSTP. This means that 'Type II error' is effectively being given different meanings by NHSTP users and critics. It becomes problematic because the critics' meaning of 'Type II error' has nothing to do with NHSTP.

6.8.   Graphical Representation of Statistical Power

A more basic question about statistical power arises when the graphical representation of the rationale of NHSTP is considered.[6] Recall from Section 3.2.4 that, although H_1 is the implication of the experimental hypothesis at the statistical level, H_1 is not used in making the statistical decision. This is the case because the experimental hypothesis is not specific enough numerically for the purpose of determining the expected mean difference. Without knowing what the mean difference is, it is not possible to specify the to-be-used sampling distribution.[7] To power analysts, this inability is responsible for psychologists' neglect of statistical power because psychologists are unable to locate precisely the H_1 distribution vis-à-vis the H_0 distribution. Consequently, psychologists cannot answer the following question:

> On the assumption that my research hypothesis is true, for the parameter of interest to me, what is the minimal value by which my research group must differ from some control group or reference value for me to be satisfied that I have established the validity of my hypothesis? (Deneberg, in Kraemer & Thiemann, 1987, p. 10)[8]

[Q6-1]

It is suggested by power analysts that Deneberg's minimal difference between μ_1 and μ_2 (e.g., 3.0) may be stipulated by convention (J. Cohen, 1987). This is the 'to-be-detected effect size' or the 'desired effect'. Furthermore, this information may be used to determine the distance between the H_0 and H_1 distributions, as illustrated in Panels A and B of Figure 6.2. However, there are difficulties if NHSTP is represented with a pair of distributions like those in Panel A or B of Figure 6.2 (also Figure 2.1 or 2.2). The difficulties become obvious if the horizontal axis of the distributions is examined more closely. Consider the index of the effect size (ES) shown in Equation [E6-1] (J. Cohen, 1987, p. 20):

$$ES = \frac{u_E - u_C}{\sigma_C} \qquad \text{[E6-1]}$$

That is, the effect size is the difference between the two population means expressed in units of the standard deviation of either the control or the experimental distribution if the two distributions are assumed to have the same standard deviation (J. Cohen, 1987, p. 20). It follows from Equation [E6-1] that the horizontal axis in Panel A or B of Figure 6.1 represents raw scores. Consequently, the pair of distributions in Panel A or B are distributions of raw scores. This raises the following problem.

If Panel A or B were a correct representation of NHSTP, statistical significance for a one-factor, two-level experiment would be determined on the basis of *two* population distributions. However, as may be recalled from Sections 2.5.3 and 2.7, the two-sample case NHSTP is based on a *single* sampling distribution (viz., the sampling distribution of differences). Moreover, the mean difference of the said sampling distribution is zero in most cases (Kirk, 1984) even when a directional H_0 is implicated (e.g., H_0: $u_1 \leq u_2$; see Section 2.2.5). In other words, the horizontal axis in Panels A or B should represent values of the difference between two means, not raw scores. The inescapable conclusion is that NHSTP is being misrepresented when statistical power is incorporated in the graphical representation, as in Panel A or B of Figure 6.2.

To power analysts, Panels A and B represent two situations in which the research manipulation is differentially efficacious. Specifically, the effect is larger in Panel B than in Panel A, as witnessed by the fact that the two distributions are farther apart in Panel B than in Panel A. At the same time, Panel B represents a more powerful test than Panel A. This is indicated by the fact that the area shaded with diagonal lines is larger in Panel B than in Panel A. That is, there is a positive relationship between the effect size and the power of the test, a feature of NHSTP already noted in Section 2.2.4. However, it has been shown that both Panels A and B are misrepresentations of NHSTP. Consequently, it becomes necessary to consider whether or not differentially efficacious research manipulations would have different impact on NHSTP. This question may be answered in the context of a one-factor, two-level experiment.

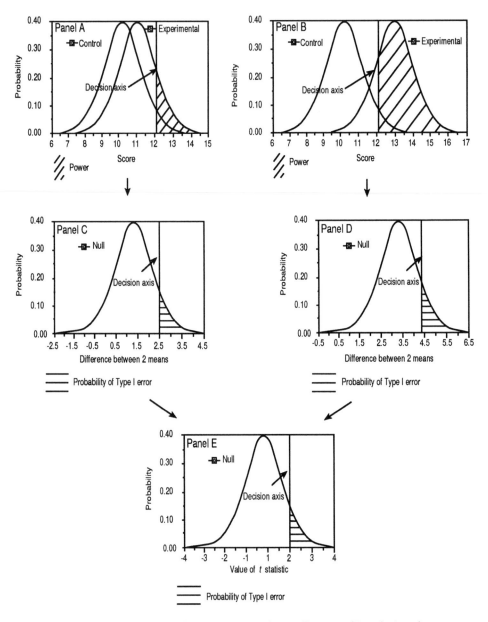

Figure 6.2   *The graphical representation of two effect sizes (Panels A and B) and the corresponding differences between two means in raw-score units (Panels C and D), as well as in standard error units (Panel E)*

6.9. Research Manipulation's Efficacy and NHSTP

The pair of population distributions in the 'small effect' situation (viz., Panel A in Figure 6.2) gives rise to the lone sampling distributions of differences depicted in Panel C of Figure 6.2. Similarly, the pair of populations distributions in the 'large effect' situation brings about another lone sampling distributions of differences, namely, the one depicted in Panel D of Figure 6.2. The two sampling distributions of differences in Panels C and D have the same standard error of the difference in the present example. However, the two sampling distributions cover different parts of the *difference between two means* continuum in raw-score units (viz., -2.5 to 4.5 versus -0.5 to 6.5).

It is now possible to return to the 'minimal difference' issue raised in Question [Q6-1] in Section 6.8. Consider the numerator used in calculating the test statistic, t. It is often written as $(\overline{X}_1 - \overline{X}_2)$. However, it is really a short-hand form for $[(\overline{X}_1 - \overline{X}_2) - (\mu_1 - \mu_2) = 0]$. The $(\mu_1 - \mu_2)$ component is left out when it is numerically equal to 0 (see Kirk, 1984). The distribution in the top panel of Figure 6.3 represents a sampling distribution of differences for a situation in which $\mu_1 - \mu_2 = 0$. That is, the mean difference of the sampling distribution of differences is zero.

Suppose that the minimal difference required between two populations means is 3.0 (or any other definite value, e.g., 1), rather than 0. The numerator now becomes $[(\overline{X}_1 - \overline{X}_2) - (\mu_1 = \mu_2) = 3.0]$. That is, the mean difference of the sampling distribution of differences implicated in NHSTP is 3.0 (or 1.0), and it is graphically represented in the bottom (or middle) panel of Figure 6.3. The three sampling distributions in the three panels of Figure 6.3 have the same standard error of differences, but different mean differences (viz., 0, 1 and 3.0). They represent the sampling distribution under H_0 in three different situations. Specifically, the bottom panel represents a more efficacious research manipulation than the one depicted in the middle or top panel.

Represented on the horizontal axis of any panel of Figure 6.3 is the range of possible values of the difference between two means. In other words, the three panels in Figure 6.3 collectively show that the difference in the efficacy of the research manipulation is represented by the spatial displacement, along the continuum of possible values of the *difference between two means*, of the sampling distribution of differences. This state of affairs is different from the impression conveyed by Panels A and B in Figure 6.2.

In carrying out NHSTP, the researcher uses a standardized form for the sampling distribution depicted in either Panel C or D of Figure 6.2 (viz., the z or t distribution; see Section 2.5.4; see also Siegel, 1956). That is, regardless of the mean difference in raw-score units, the standardized representation of the to-be-used sampling distribution of differences remains the same (viz., 0, as in Panel E in Figure 6.2). It follows that the basis of NHSTP is *not* affected by the efficacy of the research manipulation. More

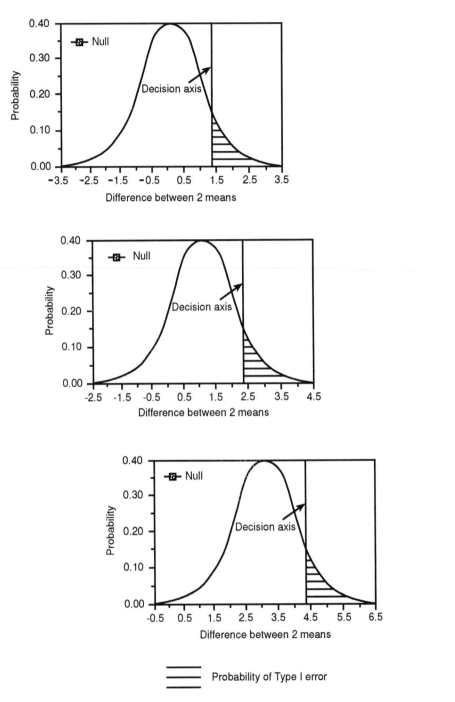

Figure 6.3   *The sampling distributions of differences in raw-score units when the mean difference is* 0 (*top panel*), 1 (*middle panel*) *and* 3 (*bottom panel*)

importantly, it is impossible to represent statistical power graphically in the sense envisaged in power analysis without misrepresenting NHSTP.

Some important points may now be made. First, no distribution based on H_1 is implicated in NHSTP (see Panel E of Figure 6.2). Second, the mean difference in raw-score units of the sampling distribution of differences reflects the theoretical difference between two population means. When expressed in terms of the raw-score unit, this difference is graphically represented by the spatial displacement of the sampling distribution on the *difference between two means* continuum (see the three panels in Figure 6.3). Third, it is not possible to represent graphically the conditional probability statistical power if the rationale of NHSTP is properly represented with a single sampling distribution of differences.

Fourth, the effect of the research manipulation (in the technical sense of the word: see Section 5.7) has no impact on NHSTP because the to-be-employed sampling distribution is standardized (e.g., in the form of the appropriate t distribution) before being used to make the 'chance versus non-chance' decision. This is the case because the location of the decision axis vis-à-vis the mean of the sampling distribution of differences remains unchanged for the same α level, regardless of what the theoretical difference between two population means is when the sampling distribution is standardized (see Panel E of Figure 6.2).

6.10. Putative Usefulness of Statistical Power Revisited

A central thesis of power analysis is that researchers should strive for as high a statistical power as their resources permit because statistical power has three important utilities. First, statistical power may be used to prevent the inadvertent rejection of a theory. Second, statistical power makes it possible to control Type II error. Third, better support for theory may be achieved by increasing statistical power. However, these three putative utilities of statistical power are debatable.

6.10.1. Prevention of the Inadvertent Rejection of Theory

Consider whether or not statistical power can prevent one from inadvertently rejecting a theory. Recall that H_0 is the complement of the implication of the experimental hypothesis at the statistical level. Suppose that statistical significance is not obtained in the experiment. The power analytic argument is that the lack of statistical significance may be due to insufficient statistical power. Typically, critics use small effect size as an indication that there is insufficient statistical power (see J. Cohen, 1965). Their solution is to increase sample size.

There are four difficulties with this power analytic argument. First, it smacks of circularity to say that the lack of statistical significance may be due to insufficient statistical power in view of [Quote 6-1] and [Quote 6-2]. This putative utility of statistical power is true by definition. The second difficulty is the debatable assumption about the relationship between the effect size and the degree of evidential support discussed in Section 5.3. Given the fact that the statistical alternative hypothesis is not the substantive hypothesis in the theory-corroboration experiment, there is no relationship between the size of the effect and the amount of evidential support available for the substantive hypothesis.

Third, the apparent attractiveness of using effect size is that it seems to be a more sophisticated quantitative index, a view endorsed by Oakes (1986). This issue raises three questions of its own. (1) Is a quantitative index more sophisticated than a qualitative one? (2) Why is a quantitative index more sophisticated than a qualitative one? (3) Is a quantitative index required at all? While the answer to (1) is taken for granted, the answer to (2) is assumed to be self-evident in power analysis. However, the crux of the matter is that Question (1) cannot be answered in absolute terms. At the same time, power analysts have not provided us with the meaningful context required for giving an affirmative answer to Question (2). A negative answer has been given to Question (3) in Section 5.3.

It has been shown that the criterion of acceptance or rejection of a research hypothesis is couched in terms of a binary statement about an ordinal relationship (e.g., partial report is superior to whole report) or a categorical relationship (e.g., Condition A is different from Condition B). In other words, the exact magnitude of the difference is not prescribed in the experimental hypothesis in either case. Consequently, a binary decision is all that is required for the purpose of assessing whether or not chance influences may be excluded. So long as a numerical difference (with or without specifying the direction of the difference, as the case may be) is established (or contradicted) in terms of a properly chosen critical value, the exact magnitude of the difference is immaterial to the theory-corroboration process. At the same time, the critical value in question is determined with reference to a lone sampling distribution of differences based on chance (i.e., H_0), not one based on H_1, let alone the research hypothesis. As for the issue of whether or not the experimental hypothesis receives support from evidence collected for that specific purpose, an appeal has to be made to both inductive and deductive logic (see Sections 4.4 and 4.5).

The fourth reason why knowing the power of the test cannot prevent the inadvertent rejection of the theory is as follows. An independent index would be required to indicate when an effect size is too small. No such index is available. Although the choice of α is arbitrary, the α level itself is well-defined in probabilistic terms (see Section 2.7). The same cannot be said of the effect size. It does not seem reassuring to appeal to the tautological assertion, 'the proposed conventions [of effect size] will be found to be reasonable by reasonable people' (J. Cohen, 1987, p. 13). An inadvertent incorrect determination of the effect size is a real possibility.

6.10.2. Statistical Power and Control of Type II Error

Consider the suggestion that statistical power may be used to control Type II error (viz., the error of accepting a false null hypothesis). This putative utility of statistical power follows from the numerical procedure used to determine power (i.e., $1 - \beta$). If one can set the power desired, one can then control the probability of Type II error (see J. Cohen, 1965, p. 98).

However, this 'control Type II error' argument raises two issues. First, for it to be relevant to NHSTP, it is necessary to show how the control over Type II error may have an impact on the outcome of NHSTP when the rationale of NHSTP does not implicate the H_1 distribution. Second, the matter is made more ill-defined by the suggestion that '[this] .80 desired power convention is offered with the hope that it will be ignored whenever an investigator can find a basis in his substantive concerns in his specific research investigation to choose a value ad hoc' (J. Cohen, 1987, p. 56). It does not seem advisable to let non-statistical factors intrude into a methodological or statistical matter.

6.10.3. Assessment of Research Result

It is suggested in power analysis that one may adjust the α value with reference to the desired power or the to-be-detected effect size. This flexibility in choosing the α value is to be justified on pragmatic or logistic grounds. This poses a problem in view of the fact that theory corroboration is a conceptual exercise to be evaluated in terms of the internal and external validity of the study.

It is true that pragmatic or logistic concerns are relevant in considering the feasibility of a research programme or applicability of research results. However, these concerns should not be dealt with at the same time as the theory–data relation is being assessed (Chow, 1991a, 1991b). To allow pragmatic and logistic considerations to intrude into the process at this stage (as recommended in power analysis) is antithetical to conceptual rigour and objectivity.

6.10.4. Converging Operations and Confidence in Substantive Hypothesis

It follows from the rationale of theory corroboration discussed in Section 4.3 that only a tentative conclusion can be drawn about the tenability of an explanatory substantive hypothesis on the basis of the data from an empirical study. If statistical power does not tell us how likely it is that a theory is true, how can we have confidence in a theory? This is a loaded question because it presupposes that our confidence in an explanatory theory is necessarily couched in probabilistic terms.

An alternative has been suggested in Section 5.8.4. That is, questions about the tenability of explanatory theories are conceptual questions about the theory–data relationship. One should not be surprised that the answer

Table 6.1　*A set of experimental hypotheses (E) predicated on some implications (I) of an explanatory theory,* T_1

	Experiment	
1	If T_1, then Implication I_{11} and E_{11} under Conditions A, B, & C	[P6-1]
2	If T_1, then Implication I_{12} and E_{12} under Conditions E, F, & G	[P6-2]
3	If T_1, then Implication I_{13} and E_{13} under Conditions A, F, & H	[P6-3]
4	If T_1, then Implication I_{14} and E_{14} under Conditions E, F, & H	[P6-4]
...	... ...	...
n	If T_1, then Implication I_{1n} and E_{1n} under Conditions A, H, & G	[P6-n]

Notes:
T_1 = the to-be-corroborated explanatory theory;
I_{1i} = implication of i of Theory T_1;
E_{1i} = the experimental hypothesis derived from Implication i of T_1;
A, B, C, etc. = test conditions for the particular experimental hypothesis.

does not (and cannot) come from statistics. Instead, the researcher's confidence in the theory comes from related experiments which collectively constitute a set of converging operations (Garner et al., 1956; Chow, 1989, 1990).

As may be recalled from Tables 5.5 and 5.6, a substantive hypothesis has more than one implication. Each of its many implications is a potential research hypothesis. For example, in the study of the psychological reality of the transformational grammar, there are many alternative versions of the [*P4.1.2*] following from [*P4.1.1*] sequence depicted in Table 4.1 (see, e.g., Miller, 1962). By the same token, each of the many exemplifications of [P4.1.2] in the said table may imply its unique experimental hypothesis in a specific experimental context. In other words, a single substantive hypothesis (e.g., [P4.1.1] in Table 4.1) begets many different experimental hypotheses (viz., tokens of [P4.1.3]). Each of these experimental hypotheses is a criterion of rejection for the substantive hypothesis. The rationale of the converging operations depicted in Table 5.6 may alternatively be illustrated schematically with Propositions [P6-1]–[P6-n] in Table 6.1.

Suppose that the to-be-corroborated explanatory, substantive hypothesis is Theory T_1. The presence of the subscript is an acknowledgment that there are other alternative explanations of the phenomenon for which T_1 (or any one of its competitors) is proposed. Proposition [P6-1] is a synopsis of the logical relations among (a) the substantive hypothesis, (b) an implication of the substantive hypothesis (viz., the research hypothesis) and (c) what the research hypothesis implies in the context of a specific set of test conditions (viz., the experimental hypothesis; see Table 3.1 and Section 3.2). With reference to Table 6.1, to conduct the first experiment is to collect data under Conditions A, B and C with a design that fulfils the formal requirement of one of Mill's (1973) inductive methods (see Section 4.5).

The experimental prescription, E_{1i}, is a criterion of rejection (if not matched by data) or of tentative acceptance (if met by data) of the substantive hypothesis by virtue of the series of three embedding syllogisms depicted in Table 4.1 or Table 4.3. Theory T_1 is accepted *in toto*, albeit tentatively, until challenged by incompatible data. Given the tentative nature of this acceptance, the onus is on the researcher, as well as the researcher's colleagues *qua* critics, to subject the substantive hypothesis to further tests. Propositions [P6-2]–[P6-n] in Table 6.1 represent further attempts to test Theory T_1 (see also Section 5.8.4).

Each of the E_{1i}'s is an independent criterion of rejection or of tentative acceptance of T_1. The theory is abandoned or modified (if justified) when any of the E_{1i}'s is not met. Confidence in the substantive hypothesis increases to the extent it can withstand the numerous attempts to falsify it (Popper, 1968a/1959, 1968b/1962). NHSTP is used in every one of the falsification attempts in exactly the same way (Chow, 1989, 1990), namely, to supply the minor premise required for the chain of three embedding syllogisms described in Section 4.3.

In short, the series of experiments schematically represented in Table 6.1 constitutes collectively the converging operations implicated in corroborating Theory T_1 (see also Table 5.6). The attempts depicted by [P6-2]–[P6-n] are not literal replications of [P6-1] at all. Hence, what is called 'converging operations' here has variously been characterized as 'conceptual replication' (Cozby, 1989), 'constructual replication' (Cook & Leviton, 1980, p. 461) or 'constructive replication' (Lykken, 1968, p. 156).[9] This is what is meant by the statement that the completion of a theory-corroboration experiment does not bring a closure to the investigation (see Row 3, 'Consequence of research', in Table 5.2).

It is instructive to examine why conducting converging operations, but not conducting literal replications, is appropriate for theory corroboration. Suppose that the researcher repeats literally the same study depicted by Proposition [P6-1] in Table 6.1 many times. However diligent and thorough, the researcher cannot be absolutely sure that there is no confounding variable in the experiment whose data support E_{11} of Theory T_1. The same doubt applies to every literal replication of the original study, regardless of who carries out the replication studies.

The situation is quite different if researchers conduct converging operations. As different experiments are implicated in the series of related studies, it becomes less likely that a confounding variable would occur repeatedly in different situations. Consequently, in the course of a series of converging operations, various confounding variables may be excluded. That is, in the course of testing different theoretical properties of the hypothetical mechanisms, various artefacts or alternative theoretical explanations are also excluded. This is the reason why psychologists do not often carry out literal replications, much to the displeasure of some critics of NHSTP (see, e.g., Bakan, 1966). In actual fact, carrying out literal replications may be misleading because they may perpetuate some confounding variables.

6.11. Summary and Conclusions

The basic concern of power analysis is how the researcher can be certain that the research data actually support the to-be-investigated theory. The difficulties with the power analytic approach to research assessment are meta-theoretical ones. There are two parts to the present critical examination of power analysis. The first criticism of power analysis is raised with the assumption that the concept statistical power is a valid one. The validity of statistical power itself is questioned in the second criticism.

Suppose that the validity of statistical power is accepted. Some of the debatable meta-theoretical features of power analysis are: (a) H_1 is incorrectly identified with the substantive hypothesis, (b) no distinction is made between NHSTP and theory corroboration, (c) the effect size or statistical power is unjustifiably assumed to have made contributions towards the evidential support for the theory, and (d) no distinction is made between statistical and non-statistical concerns. Also pertinent to the present examination of power analysis is the implicit appeal to TSD by power analysts. Apart from the fact that the similarities between NHSTP and TSD are more apparent than real, power analysis may be faulted for its debatable assumption that theory corroboration is like signal detection.

In addition to the aforementioned problems, the concept statistical power itself may be questioned because (a) it is predicated on a meaning of 'Type II error' not germane to NHSTP, and (b) its graphical representation is inconsistent with that of NHSTP. Even if these difficulties could be ignored, it is still possible to question the putative usefulness of the power of a statistical test. Statistical power cannot be the exact probability of obtaining statistical significance because statistical power is a conditional probability. Specifically, it is a probability that is meaningful *in the event that* the result is deemed not explainable in terms of chance influences. The researcher's confidence in the tenability of the explanatory theory is established by conducting a series of theoretically related converging operations. Consequently, theoretical knowledge evolves in a qualitative fashion.

Notes

1. There is an indication that some power analysts are aware of this issue, as may be seen from the comment, 'Experimental planning will frequently involve the study of the **n** [sample size] demanded by various combinations of levels of **a**, desired power, and possibly **d** [effect size], with a final choice being determined by the specific circumstances of a given research' (J. Cohen, 1987, p. 58). However, this point is not given the attention it deserves in power analysis.

2. The more exact term is signal-plus-noise event, because the signal (i.e., the additional light) is superimposed on the lit background (i.e., background noise). The noise either raises or lowers at random the observer's absolute threshold for brightness from trial to trial. These effects are assumed to be normally distributed with a mean of zero.

3. As may be recalled from Chapter 4, this is a misleading way to describe what confronts the researcher who uses NHSTP (see also Chow, 1988, 1989). Specifically, it is misleading to treat Phenomenon P itself as the to-be-investigated substantive hypothesis (see Table 3.1 and

Section 3.2). The truth of the matter is that Phenomenon P itself is often not in dispute. Research is conducted to choose among contending explanations of Phenomenon P. However, the power analytic way of talking about empirical research and NHSTP is adopted for the present discussion, even though it is a misleading mode of talking. The irony is that the difficulties of power and analysis at the conceptual level may readily be seen if the distinction is made between theory corroboration and statistical hypothesis testing, or if theory corroboration is not treated as a type of signal detection task.

4. This argument is predicated on the assumption that it is justified to treat statistical power as analogous to hit rate (see n. 2). This has a heavy Bayesian overtone (see Chapter 7).

5. This manner of speaking is predicated on the assumption that the upper panel of Figure 6.1 is a valid representation of NHSTP. However, this assumption will be questioned in Section 6.7 and thereafter.

6. It must be noted that J. Cohen (1965, 1987, 1992a, 1992b) does not use graphical representation when he introduces or discusses statistical power. Nor is graphical representation used in Kraemer and Thiemann (1987).

7. It has been shown that this state of affairs in not a liability to the purpose of corroborating the explanatory theory (see Sections 3.4 and 5.3).

8. Contrary to what is said in Chapter 3, Deneberg is treating H_1 incorrectly as the research hypothesis. This difficulty is ignored for the purpose of the present discussion.

9. Lykken (1968) distinguished between 'constructive replication' and 'multiple corroboration'. The former is similar to what is called 'converging operations' here. It seems necessary to note that Oakes (1986) seemed to reject converging operations when he said, 'Nor is it desirable to embark upon a series of experiments, each of which is analyzed by traditional tests of significance, and to attempt to make cross-experiment comparisons of the associated probabilities, obtained without regard to effect size or statistical power' (p. 73). However, Oakes's comment was one about Lykken's (1968) 'multiple corroboration'. Moreover, Oakes (1986) had in mind the situation in which each of the applications of NHSTP was based on the assumption that the substantive hypothesis had an a priori probability of .5 of being true before data analysis. This is not an assumption made when carrying out converging operations. In fact, the to-be-corroborated hypothesis is assumed to be true in every attempt to falsify it (see Section 7.7.2). Furthermore, there is no cross-experiment comparison of p's when converging operations are carried out.

7

Bayesianism

The central ideas of Bayesianism are that (1) probability refers to an individual's degree of belief in a hypothesis, (2) empirical data are collected to ascertain the inverse probability of the hypothesis of interest, and (3) the prior probability of hypotheses has an impact on research conclusions. These ideas are examined with reference to the Bayesian (i) 'personalism', (ii) sequential sampling problem, (iii) appeal to confirmation as the means to substantiate hypotheses, and (iv) relativism. As an approach to empirical research, the Bayesian procedure may be faulted for the neglect of the phenomenon-hypothesis consistency, the implication-evidence consistency, inductive conclusion validity and the rationale of theory corroboration. Bayesianism is also debatable because of its indifference to the distinctions between (a) empirical and axiomatic knowledge, and (b) objectivism and objectivity.

7.1. Introduction

In Chapter 5, the non-statistical nature of criticisms of NHSTP can be seen from the arguments in favour of using the effect size. From Chapter 6 it is learned that a motivation underlying power analysis is to adopt a direct means of ascertaining the probability of the truth of the substantive hypotheses. None the less, neither the mathematical tool used in NHSTP (i.e., the sampling distribution of the test statistic) nor the meaning of probability (viz., a relative frequency) is challenged. Bayesian critics, on the other hand, question the mathematical foundation of NHSTP.

There are two independent components to the Bayesian approach to empirical research, namely, a probability calculus and a logic of scientific inference or the rationale of empirical research (Jeffreys, 1961/1939). The arguments in favour of using Bayesian statistics do not so much constitute a rejection of NHSTP as they suggest a new approach to empirical research. Suggestive of this is Phillips's (1973) observation that statistical decisions made with Bayesian procedures are often not too different from those made with NHSTP. The differences lie more in what information to utilize and how the data are interpreted. The probability calculus found in the Bayesian approach is not an issue in the present discussion, because

[as] an exercise in mathematical formalism, Bayes' theorem is demonstrably a . . . deduction from the probability axioms. Controversy arises over the range of problems to which it may properly be applied. (Oakes, 1986, p. 99)

[P7-1]

The concern of this chapter is exclusively with the Bayesian view of the rationale of empirical research. The Bayesian theorem is first introduced with an example. The Bayesian personalist meaning of probability, as well as how it may be measured, is then described. An attempt will be made to defend treating probability as a relative frequency in view of the Bayesian *personalist probability*. Also examined are (a) the Bayesian view of the putative effects of the prior probability on research conclusions, (b) some implications of the prior and posterior probabilities from a non-Bayesian perspective, (c) the propriety of treating empirical statements as though they were formal derivations in a closed system of axiomatic rules, (d) the relativistic overtone of the Bayesian approach, and (e) the difficulty with using confirmation as the criterion of theory-acceptance. The objectivistic characteristic of Bayesianism will also be discussed.

7.2. Bayesianism

It is necessary to be circumspect in using the characterization *Bayesian* because practitioners of Bayesian analyses may disagree among themselves (Earman, 1992; Phillips, 1973). Nonetheless, a set of assumptions is shared by modern Bayesian statisticians. Earman (1992) uses the term Bayesianism to characterize these common assumptions. A Bayesian is someone who subscribes to Earman's (1992) Bayesianism in the present discussion.

For a Bayesian, 'a probability is a degree of belief held by a person about some hypothesis, event, or uncertain quantity' (Phillips, 1973, p. 13). Earman (1992) characterizes this practice as 'personalism'. The Bayesian prototype of empirical research is the sequential sampling procedure. An example of the procedure will be used to describe the Bayesian theorem. It will be shown that Bayesians do not collect data to test hypotheses: they collect data to affix subjective degrees of belief to hypotheses.

7.2.1. An Example Congenial to Bayesian Analysis

A distinction is made between the prior probability and the posterior probability of a hypothesis vis-à-vis a research project in Bayesianism. The researcher's degree of belief in the hypothesis of interest before data collection is the prior probability. The prior probability is changed into its corresponding posterior probability with Bayes's (1958/1763) theorem after data collection. The Bayesian theorem confers to Bayesianism its objectivistic and rational characteristics. Consider the following example.

Suppose that three political parties participate in an election, namely, the left-wing (L), centre (C) and right-wing (R) parties. Editor E of a daily

Table 7.1   *The accumulation of data and the conversion of the prior prob-*
ability (Prior DOB) into its corresponding posterior probability at success-
ive research stages

		Poll-data inspection period				
		1	2	3	4	5
Prior Probability	H_L	.50	.36	.31	.27	[.27]
	H_C	.60	.40	.44	.59	[.59]
	H_R	.40	.32	.24	.14	[.14]
Evidence			.30	.38	.45	.55	[.50]
Likelihood of	H_L	.38	.38	.40	.30	[.40]
Evidence*	H_C	.35	.50	.60	.75	[.60]
	H_R	.32	.35	.25	.35	[.25]
Prior Probability ×	H_L	.19	.14	.12	.08	[.11]
Likelihood	H_C	.21	.20	.26	.44	[.35]
	H_R	.13	.11	.06	.05	[.04]

Posterior Probability						
	H_L	$\frac{.19}{.53} = .36$	$\frac{.14}{.45} = .31$	$\frac{.12}{.44} = .27$	$\frac{.08}{.57} = .18$	$\left[\frac{.11}{.50} = .22\right]$
	H_C	$\frac{.21}{.53} = .40$	$\frac{.20}{.45} = .44$	$\frac{.26}{.44} = .59$	$\frac{.44}{.57} = .77$	$\left[\frac{.35}{.50} = .70\right]$
	H_R	$\frac{.13}{.53} = .32$	$\frac{.11}{.45} = .24$	$\frac{.06}{.44} = .14$	$\frac{.05}{.57} = .09$	$\left[\frac{.04}{.50} = .08\right]$

H_L = The left-wing party will form the next government. [H7-1]
H_C = The centre party will form the next government. [H7-2]
H_R = The right-wing party will form the next government. [H7-3]
Evidence = The percentage of voters polled indicate a preference for the centre party in the present example.
*Likelihood of Evidence = The probability of the evidence, given that H_C (H_L or H_R) is true.

newspaper entertains three possible hypotheses about the election outcome. They are:

H_L = The left-wing party will form the next government.   [H7-1]

H_C = The centre party will form the next government.   [H7-2]

H_R = The right-wing party will form the next government.   [H7-3]

The three 'Prior Probability' entries in Column 1 in Table 7.1 represent the respective probabilities of H_L, H_C and H_R being true to Editor E. Specifically, for Editor E, the prior probability of H_C being true is .6 in the sense that Editor E is 60% sure that H_C is true. A probability in Bayesian terms, be it prior or posterior, is the ratio of the amount of money Editor E wishes to pay to enter into a bet that H_C is true in order to receive a predetermined amount of money if H_C is true (Earman, 1992; see also Phillips's, 1973, measuring device to be described in Section 7.2.4 below). In the present example, Editor E is willing to pay $60 to bet that H_C is true with the understanding that he will receive $100 if the centre party forms the next government.

Deciding to be 75% sure about H_C (i.e., when the posterior probability is .75 or higher) before endorsing the centre party, Editor E commissions a poll to determine the percentage of prospective voters declaring a preference for the centre party. Hence, the hypothesis of interest in the example is H_C or [H7-2]. The Bayesian theorem, as well as some important concepts in Bayesianism, may be illustrated with reference to Table 7.1. As may be seen from the table, Editor E does something unusual from a non-Bayesian point of view. Instead of waiting for the poll to be completed, he examines the poll-data periodically. Each of the entries in the 'Evidence' row represents the percentage of prospective voters expressing a preference for the centre party cumulated up to the time of inspection.

Columns 1–4 represent four successive inspections. Given Editor E's '75%' criterion, the fifth inspection is not required because the posterior probability of H_C exceeds .75 (viz., 77) in the fourth inspection. For this reason, the entries in Column 5 are enclosed in square brackets. They are included here for a reason that will become obvious in Section 7.2.3. Meanwhile consider an informal illustration of a Bayesian analysis with reference to the entries in Column 1.

In Column 1 are described (1) the prior probability of each of the hypotheses before the first data inspection (viz., the 'Prior Probability' entries), (2) the percentage of voters preferring the centre party cumulated to date (i.e., the 'Evidence' entry), (3) three likelihoods assigned by Editor E to the result of the poll-data to date (i.e., the 'Likelihood of Evidence' entries), (4) the necessary computation (viz., the 'Prior Probability $\times$ Likelihood' entries), and (5) the result of the computation for the inspection period in question (i.e., the 'Posterior Probability' entries).

Specifically, (1) Editor E has different prior probabilities for [H7-1], [H7-2] and [H7-3] before the first poll is taken (viz., .5, .6 and .4, respectively[1]). (2) The result of the first poll shows that 30% of the prospective voters prefer the centre party. This is the evidence for the Bayesian analysis in question. (3) At the same time, it is Editor E's belief that the probability of obtaining the evidence is (a) .38, if Hypothesis [H7-1] is true, (b) .35, if Hypothesis [H7-2] is true, and (c) .32, if Hypothesis [H7-3] is true. (4) For each hypothesis, Editor E obtains the product of 'Prior Probability' times 'Likelihood of Evidence' (call this the 'Prior-Likelihood Product', e.g., 6 $\times$ .35 = .21 for H_C). There are three products in the example because there are three hypotheses in the universe of discourse. The three products are summed together (call this the 'Sum of Products', viz., .53 for the first inspection). (5) Editor E determines the posterior probability for each of the hypotheses by dividing the 'Probability-evidence Product' of a hypothesis by the 'Sum of Products' (e.g., .21/.53 = .40 for H_C).

The aforementioned sequence of events is repeated in each of the subsequent inspections. Of interest is the fact that the posterior probability of Inspection $n-1$ serves as the prior probability of Inspection n. For example, the three posterior probabilities of .36, .4 and .32 of the first inspection become the three respective prior probabilities of the second inspection. (See the intersection of the 'Prior Probability' rows and Column 2.) In a succinct form, the essence of the Bayesian approach to quantitative empirical research is described as follows,

> Opinions are expressed in probabilities, data are collected, and these data change the prior probabilities, through the operation of Bayes' theorem, to yield posterior probabilities. (Phillips, 1973, p. 5)

[P7-2]

7.2.2. The Bayesian Theorem

The process of revising a prior probability into its corresponding posterior probability depicted in Table 7.1 is formalized and justified with the Bayesian theorem (Jeffreys, 1961/1939). Earman's version of the theorem for multiple hypotheses is shown in the following equation (Earman, 1992, p. 34; square brackets in the denominator added):

$$Pr(H_i|E\&K) = \frac{Pr(H_i|K) \times Pr(E|H_i\&K)}{\sum_j [Pr(E|H_j\&K) \times Pr(H_j|K)]}$$

[E7-1]

where
  (a) H_i = the hypothesis of interest (i.e., H_C in the 'Editor E' example);
  (b) H_j = any hypothesis in the universe of discourse (viz., H_L, H_C or H_R in the example);
  (c) E = the new evidence for H_i (viz., 30%, 38%, 45% or 55% in the 'Evidence' row);
  (d) K = the background knowledge;
  (e) $Pr(H_i|E\&K)$ = the posterior probability of H_i (e.g., .4 in the case of H_C after the first inspection);
  (f) $Pr(H_i|K)$ = the prior probability of H_i (viz., .6 for H_C before the first inspection);
  (g) $Pr(H_j|K)$ = the prior probability of H_j (e.g., .5 for H_L before the first inspection);
  (h) $Pr(E|H_i\&K)$ = the probability of the new evidence, E, given that the hypothesis of interest is true (e.g., .35, given H_C is true for the first inspection);
  (i) $Pr(E|H_j\&K)$ = the probability of the new evidence, E, given that any hypothesis (including the one of interest) is true (e.g., .32, given H_R is true for the first inspection).

Equation [E7-1] may be informally expressed in the form of [E7-2]:

Posterior Probability

$$= \frac{\text{Prior Probability} \times \text{Likelihood of Evidence}}{\text{Sum of All (Prior Probability} \times \text{Likelihood of Evidence)}}$$

[E7-2]

In view of [P7-1], the validity of [E7-1], as a probability theorem, is granted. None the less, it is necessary to consider (a) the meanings of the prior probability and the posterior probability of a hypothesis, as well as the likelihood of evidence, at the empirical level, and (b) the claim that the form of Bayes's theorem depicted in Table 7.1 'is especially useful to the scientist who wishes to carry out experimental work in stages, for posterior opinion after one stage can serve as prior opinion to the next stage' (Phillips, 1973, p. 66). However, before embarking on this discussion, it is important to comment on the 'Editor E' example.

7.2.3. Characteristics of the Bayesian Prototype of Empirical Research

Editor E's mode of decision-making is called the 'sequential sampling procedure' (Phillips, 1973, p. 66) because of the following characteristics of the data-collection situation. First, evidential information is gathered in stages. Second, the status of the evidence is examined at intervals (e.g., a percentage in Table 7.1). Third, the evidence collected in successive stages is accumulated. Fourth, the data-collection procedure is stopped when the evidence assumes a certain value. The fourth feature may be characterized as 'self-terminating'.

These four features will be called the 'sequential-sampling features' of data-collection procedure congenial to Bayesian analyses. Moreover, it is important to emphasize that these four sequential-sampling features are not found in a typical experiment. (See, e.g., Savin and Perchonock's, 1965, experiment described in Section 3.2.3.) Furhtermore, Phillips's (1973) 'sequential sampling procedure' characterization does not reflect four other important features in the situation. For ease of exposition, these four additional features will be called the 'reflexive' features of the Bayesian data-collection procedure.

First, none of the hypotheses is proposed to explain a phenomenon which invites investigation. Instead, they are hypotheses about an uncertain event in the future. Again, this is very different from what happens in a typical theory-corroboration experiment. For example, the transformational grammar is proposed to explain the phenomenon of linguistic competence in Savin and Perchonock's (1965) experiment.

Second, the Bayesian analysis is not about the truth of hypotheses at all. Editor E does not collect data to accept or reject any of the hypotheses (e.g., H_C). Rather, Editor E is interested in the 'probabilification' of the hypothesis (Earman, 1992, p. 97). In general, 'theories are not chosen or accepted but merely probabilified' by Bayesians (Earman, 1992, p. 192). This may be seen from the fact that Editor E's decision to endorse the centre party is made before the election outcome is known.

Third, the procedure is reflexive in the sense that the termination of the data-collection procedure depends fortuitously on when the periodic data-inspection is carried out. Had the fourth inspection been delayed until Inspection 5, the evidence is 50% instead of 55%. (Hence, the 'Posterior Probability' and the 'Likelihood of Evidence' entries in Column 3 are duplicated in Column 5.) In such an event, the posterior probability for H_C is only .70, which is not sufficient for Editor E to stop the polling. On the other hand, the Bayesian sequential sampling is also an open-ended procedure in the following sense. The poll does not stop after the third inspection because the posterior probability (viz., .59 in Table 7.1) is smaller than the one desired by Editor E (viz., .75). A concomitant feature of this reflexivity is that the size of the data set is ill-defined. It is determined fortuitously by the data-collection procedure.

Fourth, the decision to stop data collection is made on the basis of a criterion not related to what is said in the hypothesis. Recall that H_c is Hypothesis [H7-2]. At the same time, Editor E's decision criterion is not the truth of H_c, but how certain Editor E is of H_c.

In contrast to the third and fourth *reflexive* features of the sequential sampling procedure, experimental psychologists do not treat their data in such a fortuitous way. Instead, the size of the data set is determined before data are collected. Specifically, experimenters adhere to their experimental plan, in which are stated, among other things, (a) the number of subjects, (b) the number of sessions a subject has to undergo, and (c) the number of trials per session. That is, there is nothing fortuitous about the size of the data set.

The four *reflexive* features of the sequential sampling procedure also explain why Bayesians talk about objectivism instead of objectivity. It is a Bayesian thesis that, regardless of the difference in the initial prior degrees of belief among different individuals, the posterior degree of belief will eventually become the same (Earman, 1992). That is, '(initially) divergent opinions will be brought more and more into agreement through the successive application of Bayes' theorem as more and more data are gathered' (Phillips, 1973, p. 14). To Bayesians, this state of affairs renders the degree of belief 'public' (Edwards et al., 1963) or 'objective' (Jeffreys, 1961/1939). 'Objectivism' is thereby achieved (Earman, 1992). The putative importance of Bayes's theorem to empirical researchers is illustrated with the suggestion that

> the scientist can design an experiment to enable him to collect data bearing on certain hypotheses which are in question, and as he gathers evidence he can stop from time to time to see if his current posterior opinions, determined by applying Bayes' theorem, are sufficiently extreme to justify stopping the experiment. (Phillips, 1973, p. 66)

[P7-3]

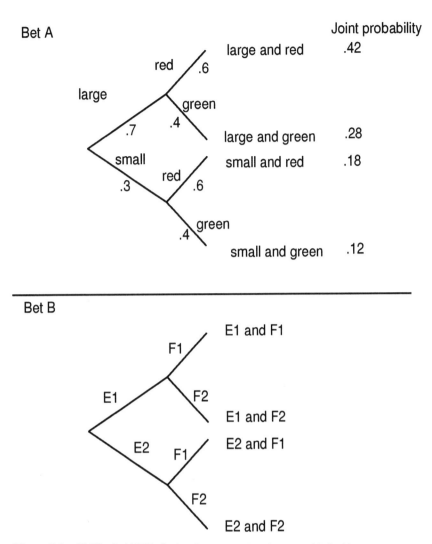

Figure 7.1 *Phillips's* (1973) *device for measuring degrees of belief* (*top panel*) *and the to-be-measured hypotheses* (*bottom panel*)

7.2.4. *How Is the Degree of Belief Measured?*

Phillips's (1973) device for measuring degrees of belief may be illustrated with reference to the two panels depicted in Figure 7.1. The measuring device is depicted in the top panel. In the bottom panel are represented four hypotheses about joint events (viz., E1 and F1, E1 and F2, E2 and F1 and E2 and F2).

Table 7.2 *Some combinations of joint probabilities for the top panel of Figure 7.1 used for measuring Individual X's degree of belief in 'E1 and F1' depicted in the bottom panel of Figure 7.1*

Row	p (large)	p (small)	p (red)	p (green)	p (large & red)	p (large & green)	p (small & red)	p (small & green)
1	.7	.3	.6	.4	.42	.28	.18	.12
2	.6	.4	.6	.4	.36	.24	.24	.16
3	.55	.45	.6	.4	.33	.22	.27	.18
4	.50	.50	.6	.4	.30	.20	.30	.20

p = probability.

Suppose that it is necessary to measure Individual X's degrees of belief in the four 'substantive' hypotheses represented in the bottom panel of the figure. Individual X's understanding of the top panel is first established with reference to two urns. Urn 1 contains 35 large balls (70%) and 15 small balls (30%). Urn 2 contains 30 red balls (60%) and 20 green balls (40%) (see the four left-most entries in Row 1 of Table 7.2). On each trial, one ball is randomly selected from each of two urns. The probabilities of obtaining each of the four joint events, 'large and red', 'large and green', 'small and red' and 'small and green' are .42, .28, .18 and .12, respectively (see the four right-most entries in Row 1 of Table 7.2).

Individual X is then asked to use the situation depicted in the top panel as a model when expressing the degree of belief in the 'substantive' hypotheses. Specifically, X is asked to choose one of two bets: Bet A is based on the top panel, and Bet B is based on the bottom panel. Given the information in the top panel, X is asked to decide whether or not Bet B (e.g., the joint event is 'E1 and F1') is preferred to Bet A (e.g., the joint event is 'large and red'). Suppose that X prefers Bet A to Bet B. This state of affairs suggests to Phillips (1973) that X's degree of belief in the joint event, 'E1 and F1', is less than .42. If X prefers Bet B to Bet A, Phillips would consider it an indication that X's degree of belief in 'E1 and F1' is higher than .42.

In either event, the probabilities of two independent events in the top panel (size and colour) are adjusted, and X is so informed, as may be seen from the four left-most entries in Row 2 of Table 7.2. Consequently, the adjusted joint probabilities will change accordingly (see the four corresponding right-most entries in the same row). Consider the case that X prefers Bet A to Bet B. Note that the new joint probabilities in the top panel are adjusted to .36, .24, .24 and .16 for 'large and red', 'large and green', 'small and red' and 'small and green', respectively (see the four right-most entries in Row 2 in Table 7.2). Individual X is again asked to indicate whether Bet A or Bet B is preferred.

Further suppose that X still prefers Bet A to Bet B. That means that X's degree of belief in 'E1 and F1' is lower than .36. The next step is to adjust the probabilities of Bet A to the entries in Row 3 in Table 7.2. Suppose that the

X now becomes indifferent as to whether Bet A or Bet B is offered. This indifference suggests to Phillips (1973) that X's degree of belief in 'E1 and F1' is .33. The precision of this measurement can be improved by changing the number of balls in some appropriate ways in the device represented by the top panel of Figure 7.1.

7.2.5. Some Reservations about the Measuring Device

Phillips's (1973) measuring device is helpful if an operational definition of probability after the fashion of Bridgman (1961/1927) is acceptable (viz., the meaning of a concept is nothing but the procedure used to measure it). However, Bridgman (1936) had shown that operationalism was not satisfactory. Moreover, the numerical values obtained with the measuring device are uninformative as to the basis of Individual X's decisions.

There is another weakness to the measuring device. Researchers are often interested in the inter-individual comparability of degrees of belief, namely, in the comparability of various degrees of belief in the same thing held by different individuals. For example, researchers may be interested in the degrees of belief in Hypothesis [H7-2] held by three individuals, X, Y and Z. For this comparability to be possible, it is necessary to assume that X, Y and Z use the same criteria with the same proficiency to assign numerical values to their respective degrees of belief. However, Phillips's (1973) device is not helpful in this regard.

Note that an individual's degree of belief in a hypothesis is subjective. At the same time, nothing in the device would prevent X, Y and Z from using different numbers to represent the same subjective degree of belief. That is, .67 for X may be subjectively the same as .49 for Y. This means effectively that X and Y are using different criteria to assign numerical values to their subjective degree of belief. The same subjective feeling may be represented by .32 for Z. On the other hand, suppose that X, Y and Z indicate their respective degrees of belief with the same numerical value (e.g., .3). It does not follow that X, Y and Z really have the same subjective degree of belief. This state of affairs may be used to explicate the realization that the rate of change in opinion is ill-defined in the Bayesian approach (Earman, 1992; Oakes, 1986). In short, Phillips's (1973) measuring device is not as useful as it may seem to be.

7.3.   Probability – Personalist or Frequentist?

The Bayesian personalist sense of probability is in contrast to the view that a probability is a relative frequency, namely, the 'frequentist' probability found in NHSTP (see, e.g., Shafer, 1993). The frequentist probability of Event E is the frequency of E relative to the total number of events in the universe of discourse. How may the personalist and frequentist approaches to probability be reconciled, if at all possible, for the empirical researcher? To this end, a critical examination of two putative difficulties with subscribing to the frequentist view of probability is instructive. It is possible to defend the *frequentist* probability.

7.3.1. Relative Frequency: Reference Class versus Reference Condition

Oakes (1986) calls the universe of discourse of a relative frequency the 'reference class'. The first putative difficulty of the *frequentist* probability arises because 'the reference class [of the associated probability implicated in NHSTP] appropriate to a situation is not always obvious' (Oakes, 1986, p. 15). Call this the 'ill-definedness' problem. At the same time, the reference class is 'the set of indefinite (usually imaginary) repetitions of the experiment which gave the result in question' (Oakes, 1986, p. 15). Call this the 'empirical impossibility' problem. That is, the *frequentist* probability may be questioned because it is said to be based on an ill-defined universe of discourse which cannot be enumerated empirically.

7.3.1.1. Issues of Reference Class

It may be shown that neither of the two issues that suggest the 'ill-definedness' of the universe of discourse of the associated probability implicated in NHSTP is germane to the validity of the associated probability as a relative frequency. In the 'reference class' case, the issue is the population to which the research conclusion may apply (see Oakes, 1986, pp. 15 and 154). The specific concern is that, when research participants are not selected randomly it is not clear of what population the sample is representative. While this issue is important, it is not one about whether or not probability in general, and the associated probability of the test statistic in particular, can be defined as a relative frequency. Oakes (1986) has raised an issue about the generality of the statistical decision.

The imaginary repetitions issue that leads to the 'ill-definedness' problem is raised in an exchange between Barnard and Savage quoted by Oakes (1986, p. 139). Barnard argues in the exchange that, if probabilities are assigned to hypotheses, it should be possible to integrate the probabilities of hypotheses to one. For such an end, it is necessary to know all possible hypotheses. Barnard argues that this necessary condition cannot be met.

Barnard's argument shows that a well-defined universe of discourse is necessary for treating probability as a relative frequency. However, it is not an objection to treating the associated probability of the test statistic as a relative frequency because Barnard is not questioning the universe of discourse of the *frequentist* probability. Instead, Barnard's argument is one about a difficulty with the Bayesian practice of assigning probabilities to substantive hypotheses. Barnard is arguing against Bayesianism. In using Barnard's argument to illustrate the 'ill-definedness' problem, Oakes (1986) treats the collection of possible hypotheses for the set of data in question as the universe of discourse of the associated probability. However, it is debatable to do so for the following reason.

Oakes (1986) also makes reference to Bakan (1966) when the 'ill-definedness' problem is raised, even though Bakan (1966) is not concerned with the universe of discourse of the test statistic. Instead, Bakan is concerned with the inference warranted by a statistically significant result. His point is that, knowing that a sample of normal patients have a mean score significantly different from that of a sample of schizophrenics, one may infer that the two respective population means are different if the research participants are randomly selected (viz., Bakan's, 1966, induction to the aggregates). That is, it is warranted to make statistical inferences about data (see also Phillips, 1973, p. 233). However, one is not thereby warranted to say that all normal individuals are different from all schizophrenic individuals (viz., Bakan's, 1966, induction to the general). In other words, the said statistical significant result does not warrant one to say anything about the nature of schizophrenia. Bakan's point is that statistical inferences are not about substantive hypotheses (see Phillips, 1973, p. 233). Oakes (1986) has not demonstrated that the universe of discourse of the *frequentist* probability is untenable because neither Barnard's nor Bakan's argument is concerned with the universe of discourse of the associated probability.

7.3.1.2. Reference Condition – A Rejoinder to Empirical Impossibility It is possible to appreciate that the associated probability of the test statistic is a relative frequency if the universe of discourse is characterized as the 'reference condition' instead of the 'reference class'. Suggestive of this are (a) Bakan's (1966) distinction between induction to the aggregate and induction to the general, (b) Phillips's (1973) observation that statistical 'inferences [are applied] to measurements, not attributes' (p. 233), and (c) what is said in Sections 2.5 and 2.6.2 above. Furthermore, the 'reference condition' characterization may also be used as a rejoinder to the 'empirical impossibility' problem identified above.

The reference condition of the associated probability of a test statistic may be illustrated with a recapitulation of the sampling distribution of differences as follows:

Step 1: Suppose that the two populations of interest are identical in that they both consist of the scores 1, 2, 3, 4, 5, 6, 7 and 8 that are distributed in the way described in Table 2.6.

Step 2: Draw a random sample of 15 from each of the two populations, and calculate their means (i.e., $\overline{X}_1$ and $\overline{X}_2$).

Step 3: Calculate the difference between the two sample means, namely, $(\overline{X}_1 - \overline{X}_2)$.

Step 4: Return the two samples to their respective populations.

Step 5: Repeat Steps 2–4 T times.

Step 6: Let x be the number of times in which the difference between the means of the two random samples, $(\overline{X}_1 - \overline{X}_2)$, is 0.33 or more extreme.

Step 7: The probability of having $(\overline{X}_1 - \overline{X}_2) = 0.33$ or more extreme is the limiting value of x/T as T approaches infinity.

The definition of probability in Step 7 is characterized as frequentist (Earman, 1992; Gigerenzer, 1993) because probability is expressed as a ratio between two frequencies (viz., x and T in the present example). It may be seen that the relative frequency of a difference of .33 or differences more extreme values in Step 7 is predicated on the data-collection procedure (viz., x trials out of a total of T random trials of selection of two samples of 15 each from the respective populations which have the same mean). In other words, underlying probability statements in empirical research, there is not a reference class, but rather a reference condition (viz., the data-collection condition).

It may also be seen that the concern with the 'empirical impossibility' problem is not warranted. First, far from being 'imaginary', the reference condition is specified by the design of the experiment, as well as the data-collection procedure (see Section 2.6.2). Second, it is not necessary to establish empirically the limiting value of x/T in Step 7. Instead, it is available by virtue of the Central Limit Theorem. That is, the sampling distribution that renders the research data probabilistic is a theoretical distribution warranted by the Central Limit Theorem. As the Central Limit Theorem is a well-established theorem in mathematics, it renders the 'imaginary' characterization debatable.

The 'indefinite' characterization gives the impression of being ill-defined. However, it is misleading because the sampling distribution is fully specified with the appropriate statistics obtained from the relevant samples (and, hence, well-defined). For example, the sampling distribution of differences is fully specified if the mean difference, $(\overline{X}_1 - \overline{X}_2)$, the sizes of the two samples (i.e., n_1 and n_2), the correlation between the data from the two samples (in the case of two related samples) and the respective standard deviations (s_1 and s_2) of the two samples are known. Hence, it is more appropriate to say that the relative frequency of the test statistic is mathematically determined in the context of a well-defined data-collection situation.

7.3.2. *Bayes's Objective Probability*

The second, and a more telling, difficulty with the anti-frequentist position is the fact that Bayes (1958/1763) himself actually recognized the *frequentist* probability, as well as the need to postulate a hypothetical propensity or mechanism. Earman (1992) notes that Bayes recognized a tendency in chance setups to yield stable frequencies. The tendency in question is the result of the physical propensities in objects (i.e., presumably as a result of the unobservable hypothetical structures or mechanisms discussed in Chapter 3).

The stable frequencies so obtained are objective probabilities in the sense that they are not dependent on an individual's sense of certitude (Earman, 1992). Moreover, an objective probability is dependent on a ratio between two frequencies. This meaning of probability is frequentist par excellence. Important for the present discussion is the fact that the objective probability is assigned to data, not the substantive hypothesis, in Bayes's (1958/1763) account (see the Appendix to this chapter).

To conclude, it is the present argument that probability statements in empirical research are statements about research data. This way of looking at the probabilistic character of NHSTP is consistent with Bayes's objective probability. It is, hence, reasonable to apply the probability calculus when probability is used in the frequentist sense to talk about data.

7.3.3. Linkage between the Two Meanings of 'Probability'

The Bayesian *personalist* probability, on the other hand, is the individual's degree of belief in the hypothesis. Strictly speaking, it is inappropriate to use the frequentist probability calculus with the Bayesian *personalist* probability for the reasons that (a) the individual's degree of belief and research data belong to different domains, and (b) the concept *chance* plays no well-defined role, if it plays any role at all, in the quantification of the degree of belief (see Jeffreys, 1961/1939).

At the same time, it is necessary to provide a rationale for linking the *frequentist* and *personalist* meanings of probability if the probability calculus is to be applied to the degree of belief in a hypothesis. This necessity is emphasized because it has been noted that 'for Bayes probability *qua* degree of belief is assigned to events' (Earman, 1992, p. 9), and that '[when] dealing with events, you will often find that your degrees of belief are identical to the proportion of elementary events in an event class' (Phillips, 1973, p. 16). However, entertaining an implicit assumption, or making an assertion about, the linkage between the *frequentist* and *personalist* meanings of probability is not providing a justification or an explication. Symptomatic of this difficulty is the possibility of making the two conflicting statements, [P7-4] and [P7-5] at the same time:

> the quantitative approach in terms of degrees of belief [is] regimented according to the principles of the probability calculus. (Earman, 1992, p. 33)
>
> [P7-4]

> one could try to exploit the interpretation of probability as degree of belief as a means of getting a justification for the probability axioms (Earman, 1992, p. 38)
>
> [P7-5]

In short, there are good reasons to distinguish between probability and degree of belief. Hence, to distinguish between the personalist and frequentist meanings of probability, the phrase *degree of belief* will be used to refer the Bayesian personalist meaning of probability in subsequent discussion. The term *probability* will be restricted to its frequentist sense.

7.4. A Non-Bayesian View of the Prior Degree of Belief

In the Bayesian account, the actual contribution of research data to changing the researcher's degree of belief in a hypothesis depends, among other things, on the degree of belief in the hypothesis before data collection (i.e., the prior degree of belief). The putative impact of the prior degree of belief on research conclusions envisaged in Bayesianism will be discussed in Section 7.7. Meanwhile, the concept prior degree of belief itself is examined.

Suppose that Researcher R entertains the same degree of belief in Hypotheses H_1 and H_2 before the research project. For Jeffreys (1961/1939), R is *not* saying that H_1 is as likely to be true as H_2; R is simply showing an 'equal distribution of ignorance' (Jeffreys, 1961/1939, p. 34) between H_1 and H_2. To see the difficulties with Jeffreys's position, it is important to distinguish between saying (a) that R is as uncertain about H_1 as R is uncertain about H_2 and (b) that H_1 is as satisfactory as H_2 in terms of the phenomenon-hypothesis *consistency* (see Section 3.2.1 above). Jeffreys (1961/1939) seems not to have distinguished between the two.

None the less, it is instructive to examine what it means for R to say that R is uncertain about H_1 or H_2. It means that R has some reservations about the tenability of H_1. In other words, any 'less than 1' prior degree of belief in H_1 in Bayesianism is actually a concern about the tenability of H_1 in the theory-corroboration perspective.[2] At the same time, questions about the tenability of H_1 are meaningful only when there are well-defined criteria for assessing the tenability in question. Seen in this light, something is amiss in Jeffreys's (1961/1939) indifference to the phenomenon for which the hypothesis is postulated. Now consider what it means to say that H_1 is as satisfactory as H_2 from a non-Bayesian perspective.

7.4.1. A Non-Bayesian View of the Prior Probability

Hypotheses are not postulated in the void. Non-Bayesian empirical researchers propose hypotheses to account for phenomena, which precedes attempts to explain them. Hence, R's assertion that R has the same degree of belief in H_1 as in H_2 is of theoretical interest only if the assertion describes R's judgement about the interrelationships among H_1, H_2 and the phenomenon for which the hypotheses are postulated (e.g., Phenomenon P). If Phenomenon P is recognized, R's statement may mean that H_1 is as consistent with P as H_2 is consistent with P. This is the phenomenon-hypothesis consistency or prior data-hypothesis consistency mentioned in Section 3.2.1.

Why do questions about the phenomenon-hypothesis consistency arise in the theory-corroboration perspective, but not in Bayesianism? The phenomenon-hypothesis consistency issue does not arise in Editor E's situation because there is no pre-existing phenomenon to describe or explain. In other words, Bayesians do not consider the phenomenon-hypothesis consistency because they treat the sequential sampling procedure as the prototype of empirical research. However, the sequential sampling procedure is, in fact, unlike any of the kinds of empirical research identified in Chapter 3.

The phenomenon-hypothesis consistency is an iteration of the 'phenomenon→hypothesis' sequence identified in Section 3.2.1. The exact nature of the 'phenomenon' component of the 'phenomenon→hypothesis' sequence changes as the research progresses. As may be recalled from Tables 5.5 and 5.6, the phenomenon-hypothesis consistency is assessed with an increasingly more specific empirical criterion as research progresses. Indicative of this is the fact that the to-be-corroborated implication at any stage of the research has to be consistent with more research data in addition to the to-be-explained phenomenon. (For example, compare the conjunction, $P+D_1+D_2$, to P in the 'Phenomenon/Prior Data' column in Table 5.6.) Consequently, the various implications (viz., $I_1 \ldots, I_t$) may be qualitatively different theoretical assertions.

Consider the prior data for Experiment 3 with reference to Table 5.6. The prior data consist of P, D_1 and D_2. Questions about the phenomenon-hypothesis consistency at the beginning of Experiment 3 are, in fact, questions as to whether or not D_2 collected in Experiment 2 is unambiguously consistent with the theoretical implication of Experiment 2 (viz., I_2). These questions are seldom about the statistical conclusion validity of Experiment 2. More often than not, they are questions about the inductive conclusion validity or hypothesis validity of Experiment 2. These questions do not arise for Bayesians because they do not consider objectivity possible or germane to the purpose of research.

In sum, the multiple occasions of data collection in a series of converging operations in theory corroboration are different from the occasions of periodic data inspection envisaged in the Bayesian approach. Bayes (1958/1763) was distracted from those considerations important to theory-corroboration experimental psychologists by the type of research problem he had in mind (viz., the sequential sampling procedure). None the less, Bayesianism may be faulted to the extent that the phenomenon-hypothesis consistency is neglected. As the Bayesian theorem is based on the sequential sampling procedure, the propriety of applying the theorem to situations other than the sequential sampling procedure situation may be questioned, the assertion [P7-3] notwithstanding (see Section 7.2.3 above).

7.4.2. *The Impact of the Prior Degree of Belief*

To Bayesians, the prior degree of belief has an impact on the posterior degree belief. Specifically, the prior degree of belief actually determines the extent to which the evidence contributes towards the posterior degree of belief. In contrast, non-Bayesians do not specify their prior degree of belief before conducting the research. Nor do they take the prior degree of belief into account when they draw their research conclusions (see Phillips, 1973, pp. 72–8, 328). In view of the two conflicting views (viz., to ignore or not to ignore the prior degree of belief), it becomes necessary to consider more closely the putative effects of the prior degree of belief on the posterior degree of belief envisaged by Bayesians.

Suppose that a clinician entertains only two explanations of a behavioural problem (BP), namely, an unhealthy diet or inappropriate social conditioning. Further suppose that the clinician has to decide which of the two hypotheses better explains the symptoms of the boy sent to the clinic. These two alternatives are listed in two rows of each of the four panels depicted in Table 7.3.

Panels (1) and (2) of Table 7.3 represent the situation in which the clinician has no preference for either hypothesis before seeing the boy. (Hence, the prior probability of either hypothesis is .5 in the 'Prior DOB' column.) Panels (3) and (4) of Table 7.3 represent the situation in which the clinician has a higher prior degree of belief in the Diet hypothesis (viz., the prior degree of belief of .78) than in the Socialization hypothesis (viz., the prior degree of belief is .22; see the 'Prior DOB' column).

As a diagnostic tool, the clinician uses the attention span task. The boy is shown a very brief visual display (say, a duration of .1 second) followed by another stimulus that serves as a mask, and asked to recall all digits shown in the display. The task may begin with a display of Size 1 (i.e., one digit in the display). The display size increases by one digit from trial to trial so long as the boy recalls all the digits correctly. The test stops as soon as the boy fails to report all the digits in the display correctly. Suppose that the boy fails at Size 5. The boy's attention span is four digits. Further suppose that an attention span of fewer than four digits is called 'short'. It is a 'long' attention span if the boy can recall four or more digits.

The entries in the 'Likelihood' columns are conditional probabilities, namely, the probability of a boy having a short attention span (or a long attention span), given that the child is on the unhealthy diet (or having been improperly socialized). The probability of the boy having a short attention span, given the Diet hypothesis, is .73; but it is .30, given the Socialization hypothesis (see Panel 1 of Table 7.3). The probability of the boy having a long attention span, given the Diet hypothesis, is .27; but it is .70, given the Socialization hypothesis (see Panel 2). The likelihood information is obtained from an earlier study.

Suppose that the clinician uses the boy's attention span to make the diagnosis. Panels (1) and (3) represent the situation in which the boy's attention span is short. The boy's attention span is long in Panels (2) and (4). To Phillips (1973), the clinician, *qua* a researcher, also uses the boy's attention span to assess the clinician's own degree of belief in the two hypothesis. The computations in Table 7.3 are based on Equation [E7-1] introduced in Section 7.2.2.

The contribution of the evidence to the posterior degree of belief in either of the two hypotheses is demonstrated by Phillips (1973) as follows. Consider Panels (1) and (2) together. The posterior degree in belief in Diet is higher than that in Socialization when the boy's attention span is short

Table 7.3   *Four applications of Bayes's (1958/1763) theorem*

The boy's attention span	Hypothesis	Prior DOB*	Likelihood [p(Data|H)]	Prior DOB × Likelihood	Posterior DOB
(1) Short‡	Diet	.5	.73	.365	$\frac{.365}{.515} = .709$
	Socialization	.5	.30	.150	$\frac{.15}{.515} = .291$
(2) Long†	Diet	.5	.27	.135	$\frac{.135}{.485} = .278$
	Socialization	.5	.70	.350	$\frac{.350}{.485} = .722$
(3) Short	Diet	.78	.73	.569	$\frac{.569}{.635} = .896$
	Socialization	.22	.30	.066	$\frac{.066}{.635} = .104$
(4) Long	Diet	.78	.27	.211	$\frac{.211}{.365} = .578$
	Socialization	.22	.70	.154	$\frac{.154}{.365} = .422$

*DOB = degree of belief.
‡Short = attention span is fewer than 4 digits
†Long = attention span is 4 or more digits

(viz., .709 versus .291 in Panel 1). The posterior probability is reversed when the boy's attention span is long (i.e., .278 versus .722 in Panel 2). That is, when the prior degree of belief in both hypothesis is the same (viz., .5), the clinician would favour the Diet hypothesis if the boy's attention span is short, but the Socialization hypothesis if his attention span is long. That is, the clinician's choice between Diet and Socialization is determined by the evidence (viz., the boy's attention span) when the clinician has no preference for either hypothesis.

In Phillips's (1973) view, the impact of the prior probability is more readily seen when Panels (3) and (4) are considered. The prior degrees of belief in Diet and Socialization are .78 and .22, respectively. Unlike Panels (1) and (2), the posterior degree of belief in Diet is higher than the posterior degree of belief in Socialization, regardless of what the boy's attention span is. In other words, the clinician would favour the Diet hypothesis even though the boy's attention span is long. This state of affairs suggests to Phillips (1973) that, even though the evidence contributes to the change of opinion about the hypotheses, its contribution may be overwhelmed by that of the prior degree of belief. Hence, Phillips (1973) concludes that the prior degree of belief should not be ignored. Specifically,

> Whether or not the [boy's short attention span] tells the clinician something about the [boy] . . . the information conveyed by the test results is far less than that shown in the prior probabilities. In this case the prior probabilities swamp out the information in the test, so that the posterior probabilities are determined more by the priors than by the likelihoods. The extra information given by the test does not change the prior probabilities enough to warrant giving the [attention span] test. (Phillips, 1973, p. 74)
>
> [P7-6]

7.4.3.   The Impact of the Prior Degree of Belief Revisited

What is said in [P7-6] is, in fact, the last thing to which theory-corroboration researchers should subscribe. Recall the differences between the theory-corroboration experiment (see Section 3.2) and the clinical experiment (see Section 3.6.2). The question for the theory-corroboration experiment is whether the observed attention span warrants accepting the Diet or the Socialization hypothesis. In contrast, the tenability of the two hypotheses must be assumed in the case of the clinical experiment. The question is which of the two accepted hypotheses is applicable to the boy. The important differences between the theory-corroboration and the clinical questions have not been made in [P7-6]. This difficulty of the Bayesian approach may also be phrased in the following way.

To Phillips (1973), the clinician is making a diagnosis and adjusting the clinician's own degrees of belief in the two hypotheses at the same time. This is unacceptable for two reasons. First, how is the clinician justified to use the hypotheses in making the diagnosis if there are reservations about the tenability of the hypotheses (i.e., if the prior degree of belief is less than 1)? Assuming the clinician and researcher roles simultaneously, the clinician is confusing the tenability of an explanatory hypothesis with the applicability (to a particular case) of the explanatory hypothesis whose tenability is taken for granted (see Section 3.6.2).

Second, having a short or long attention span is assumed to be a symptom of the said behavioural problem, BP. In this example, either symptom is consistent with both the Diet and Socialization hypotheses. While this symptom may be acceptable as a diagnostic index to be used to determine whether or not there is an instance of BP, it should never be used as the evidence for deciding the tenability or applicability of the two alternative explanations of the behavioural problem in question. The clinician should have chosen a measure other than attention span if the clinician wishes to ascertain the tenability of the two hypotheses. Moreover, the new measure must be such that (a) it does not implicate the behavioural problem, BP, and (b) it is consistent with one, but not both, of the two hypotheses (see also Section 7.8 below).

An important assumption underlying Equation [E7-1] or [E7-2] in Section

7.2.2 is that the set of hypotheses in question are mutually exclusive and collectively exhaustive (Earman, 1992). This stipulation is easy to fulfil at the level of an axiomatic system. (See Phillips's, 1973, imaginary conversation in which he convinces a scientist of the validity of the Bayesian theorem, pp. 67–70.) However, it is a debatable assumption at the empirical level. For example, there is no self-evident reason why the Diet and Socialization hypotheses about the behavioural problem BP should be mutually exclusive and exhaustive alternatives. In other words, it is misleading to use the exercise in Table 7.3 to illustrate the importance of the prior degree of belief on research conclusion (see also Section 7.7.2 below).

7.4.4. Some Meta-theoretical Reservations

Recall that the Bayesian posterior probability is the product of prior probability times likelihood. Consequently, the validity the posterior degree of belief is dependent on the validity of the likelihood in question. From where is the likelihood obtained? Suppose that it is obtained from an earlier study. (Call it the 'source' study.) The difficulty is that the researcher would have to take for granted that the source study is valid and the measurements in the source study are accurate. Is it advisable to base the validity of one study on the validity of another study?

The clinician in the present example may have derived the prior degree of belief from many sources (e.g., hospital and clinic records, research literature, and the like). However, the overriding mode of arriving at the degree of belief is by appealing to 'certain self-evident principles (axioms) of rationality' (Phillips, 1973, p. 75). It is not clear how the principles of rationality envisaged by Phillips may be identified. To require that these principles be 'self-evident' is not satisfactory.

A telling difficulty with the Bayesian attitude towards the posterior degree of belief may be seen from the conclusion Phillips (1973) might draw from Table 7.3. Use Panel 1 as an example. The conclusion is that the clinician is 70.9% sure that the boy suffers from behavioural problem BP because of his unhealthy diet and 29.1% sure that the boy suffers from inappropriate socialization (see Phillips, 1973, p. 62). However, Bayesians do not ask the important question as to whether or not either hypothesis is correct. The clinician's certitude has nothing to do with the tenability of the hypotheses.

7.5. Axiomatic Probability Theory versus Empirical Research

It has been noted that the phenomenon-hypothesis consistency issue is neglected in Bayesianism. However, Bayesians are very attentive to a different kind of consistency, namely, the consistency among beliefs. This may be understood better if it is made explicit that Bayesians treat the probability calculus as the formalization of inductive reasoning. A good example is Jeffreys's (1961/1939) book, *Theory of Probability*.

Jeffreys (1961/1939) attempted to propose an extended logic, namely, a logic which includes both inductive logic and deductive logic. His objectives were (a) to provide the inductive rationale of making generalization, and (b) to provide the inferential facilities normally rendered possible by deductive logic.[3] He had in mind a system of axiomatic rules in which whatever could be said about an individual's beliefs could be deduced from a limited set of axioms. Hence, internal consistency is important because anything can be derived from a pair of inconsistent premisses (Copi, 1965, 1982). In fact, Bayesians are concerned with the consistency of among one's various beliefs (see Phillips, 1973, p. 69). In Oakes's (1986) words,

> whereas in many branches of science beliefs or theories are tested against reality, a corresponding procedure for Bayesians would present extremely embarrassing philosophical problems. The discipline, therefore, that Bayesian inference requires from its adherents is not that their beliefs (probabilities) should correspond to reality but that they should be consistent (or 'cohere'). (Oakes, 1986, p. 135)

This is acceptable if the theory of probability is used as a closed, axiomatic system. It is not appropriate if the theory of probability is meant to be more than an axiomatic system. In actual fact, Jeffreys (1961/1939) treats his theory of probability as a general rule of empirical research. Someone using such a closed, formalistic system to characterize empirical research is a 'logically omniscient scientist' (Earman, 1992).[4]

However, empirical research is, by definition, an open system in that not all empirical statements can be deduced from a limited set of self-evident axioms. Moreover, 'flesh-and-blood scientists are not logically omniscient . . . [because they may fail] to recognize logical implications . . . [or fail to] perceive what alternative theories lie in the space of possibilities' (Earman, 1992, p. 168). In short, it is not clear how the fundamental differences between an open, empirical system and a closed, axiomatic system may be reconciled.

What is important is that the internal consistency important to a system of axiomatic rules is neither the phenomenon-hypothesis or prior data-implication consistency required of empirical hypotheses (see Section 3.2.1 and Section 7.4.1 above) nor the hypothesis-evidence or hypothesis-evidential data consistency required of empirical research (see Sections 3.2.4 and 7.4.1 above). That is, axiomatic rules are not appropriate for an empirical discipline, such as psychology. (See Cotton's, 1955, and Koch's 1944, 1954, discussion of the Hullian system; see also Turner, 1967.)

For example, how is it possible for empirical researchers to have new insights into familiar phenomena if researchers entertain only ideas that are consistent with the pre-existing system of beliefs? On the contrary, there is nothing inappropriate in entertaining conjectures that are radically different from current beliefs. Bold conjectures (even when they are contrary to established beliefs) may lead to better understanding if they survive rigorous attempts to falsify them along the lines described in Chapters 3 and 4 (Popper, 1968a/1959, 1968b/1962; see n. 6 in Chapter 5).

A moral of this story is that, apart form the sequential sampling procedure situation described in Section 7.2.1 (Bayes, 1958/1763; Edwards, et al., 1963; Phillips, 1973), the initial concern of the empirical research is not (and should not be) the researcher's degree of belief in the hypothesis before conducting the research. Nor should it be whether or not the hypothesis is consistent with the rest of the belief system. Instead, the initial concern should be whether or not the hypothesis is consistent with the to-be-studied phenomenon. The next concern is whether or not the theoretical mechanisms postulated in the new hypothesis are well-defined enough for rigorous empirical testing.

7.6. Bayesian Objectivism and Rationality Revisited

Important to Bayesians is objectivism, not objectivity. By objectivism is meant that, to Bayesians, it is possible to achieve inter-subjective agreement on the degree of belief in the hypothesis with repeated applications of the Bayesian theorem. This agreement is assumed to be rational because it is not the result of just some historical or sociological factors. The agreement has a 'justificatory character' (Earman, 1992, p. 138).

It may be pointed out that the consensus is reached in the Bayesian account only when a sufficient amount of data has been collected. What amount of data is deemed sufficient? The answer does not inspire confidence for two reasons. First, as has been noted before, the rate of revision of the posterior degree of belief is not known (see Oakes, 1986). One is not able to tell when sufficient data are available even if one knows what the prior degree of belief in the hypothesis is. Second, there is no guideline as to when data collection can (or should) stop, which is independent of the fact that a consensus (in the posterior degree of belief) has been reached. Consequently, it seems that we are confronted with the following circularity:

[a] Consensus will be reached when sufficient data are available.

[b] Sufficient data have been collected when a consensus is reached.

Two more points may be noted about this objectivism. First, while the objectivism may well be achievable at the mathematical level, the assessment of empirical findings implicates more than arriving at the same numerical value for different individuals' personalist degrees of belief. This point needs to be made. Although Bayesians are concerned with objectivism rather than objectivity, some Bayesians (e.g., Earman, 1992) subscribe to the Kuhnian view that 'it is the community of specialists rather than its individual members that makes the effective decision' (Kuhn, 1970, p. 200). Second, as may be recalled from Section 7.2.4, Phillips (1973) has not demonstrated the possibility of the inter-individual comparability of the degree of belief. That is, an inter-individual agreement in a numerical value does not mean that the individuals' degrees of belief are justifiably comparable.

Objectivity, as it is understood by non-Bayesians or non-Kuhnians, refers to inter-individual agreement with reference to a pre-determined set of impersonal, non-circular criteria. In other words, the Kuhnian inter-individual agreement is, at best, the necessary requirement. It is by no means sufficient for the purpose of defining objectivity. The agreement must be based on criteria not susceptible to personal whims, even though it may be the whims of experts. At the same time, it has been shown that such an impersonal criterion for theory corroboration is available in the form of (a) theoretical implications expressed as the experimental prescriptions at the conceptual level (i.e., Section 4.3) and (b) the mechanical decision rule, NHSTP, at the statistical level (Section 2.2).

7.7.   Prior Degree of Belief – Personalistic and Relativistic

Bayesians may fault the non-Bayesian objectivity by pointing out, correctly, that (a) different researchers may (and, more often than not, do) prefer different hypotheses, (b) researchers cannot overcome the influence of their prior degrees of belief in the hypothesis (as claimed by Phillips, 1973), (c) the prior degree of belief may overwhelm the new evidence (see Section 7.4.2 above), and (d) there is no impersonal, non-circular criteria of objectivity (Gergen, 1991). That is, Bayesians seem to suggest that (b)–(d) are inevitable in view of (a). This suggestion may be examined with reference to a proposition with a Bayesian overtone:

> **Researcher R's degree of belief in Hypothesis H is .75, whereas**
> **Researcher S's degree of belief in Hypothesis H is .37.**
>
> [P7-7]

Proposition [P7-7] means that the two researchers have different faith in the hypothesis before conducting the research about Hypothesis H. Hence, as has been noted before, Bayesians bemoan the fact that non-Bayesian researchers do not take into account this initial difference in the prior degree of belief in the hypothesis when non-Bayesian researchers conduct research (see, e.g., Phillips, 1973). In emphasizing the importance of the prior degree of belief in the hypothesis in the way they do, Bayesians seem to suggest that, since Researchers R and S have different degrees of belief in Hypothesis H before conducting the research, Researchers R and S might (a) conduct the research differently or (b) be warranted to interpret the data differently.

That the prior degree of belief is important in the manner implied in Bayesianism seems convincing in view of the general acceptance of the various social influences on experimentation collectively called the social psychology of the experimental experiment by Orne (1962) and Rosenthal (1963) or SPOPE, for short, by Chow (1987a, 1992a). The SPOPE factors are experimenter effects, particularly, the experimenter expectancy effects (Rosenthal, 1976; Rosenthal & Fode, 1963a, 1963b), subject effects (Rosenthal & Rosnow, 1975) and demand characteristics (Orne, 1962, 1969). Hence, to accept the Bayesian characterization of the importance of the prior degree of belief is to concede that it is not possible to have objectivity based on impersonal, impartial criteria. This concession may lead one to accept Gergen's (1991) argument that theory acceptance is a matter of rhetoric, not a decision based on objective data because there are no non-circular data.

7.7.1. Relativism Revisited

Given its relativistic overtone, Bayesianism loses its persuasiveness to the extent that its relativistic overtone is problematic; and the relativistic overtone is indeed questionable. Gergen (1991) is correct in saying that all observations are theory-dependent. However, it does not follow that all observations are necessarily circular. This is the case because there is more than one possible theory. Specifically, Theory X may be tested with data whose identity depends on a different theory (say, Theory Y), and Theory X belongs to a domain radically different from the domain of Theory Y (Chow, 1992b).

This point may be explicated by recalling Savin and Perchonock's (1965) study of the transformational grammar as the basis of our linguistic competence (see Section 3.2.3). The dependent variable is the number of extra words recalled after the verbatim recall of a sentence. It is true that the experimental expectation of the study is dependent on the to-be-studied transformational grammar. However, the identity of the subject's utterances in the experiment (i.e., whether or not what the subjects uttered in the recall phase were actually words), and the correctness of the subject's responses (i.e., whether the words were in the list of extra words presented on the trial) are dependent neither on the transformational grammar nor on any other grammar in competition with the transformational grammar. That is to say, the fact that Savin and Perchonock (1965) subscribed to the transformational grammar did not affect (a) the identity of the dependent variable, or (b) how the responses were tabulated.

In short, the putative circularity alluded to by Gergen (1991) is more apparent than real for the following reason. The theory to which experimental observations owe their identity or meaning (call it the 'data-theory') is different from, as well as independent of, the to-be-corroborated theory (Chow, 1992b). An implication of Chow's (1987d) objection to the 'ecological validity' argument is that the data-theory *should be different from* the to-be-corroborated theory, a point to be developed further in Section 7.8 below.

As far as the SPOPE arguments are concerned, data suggestive of the experimenter expectancy effects are not satisfactory because Rosenthal and Fode (1963a, 1963b) conducted some experiments when meta-experiments were called for (viz., experiments about experimentation). At the same time, meta-experimental data do not support the experimenter expectancy effects (Chow, 1994).

The putative subject effects are questionable because the necessary distinction is not made between group difference and the effects of the group difference in studies often cited in support of the subject effects (Chow, 1992a). Only the former is reported (e.g., Goldstein, Rosnow, Goodstadt, & Suls, 1972) when the latter is required for substantiating the claim of subject effects.

It has been observed by Berkowitz and Donnerstein (1982) that there is no evidence for the putative demand characteristics in experimental studies because the often cited 'evidence' in support of demand characteristics consists of nothing but some anecdotes of good will (Chow, 1992a). Although the anecdotes reported by Orne (1969) and Orne and Evans (1965) are interesting, they are not experimental data because the anecdotes were collected in situations which did not satisfy the formal requirement of the inductive rules described in Chapter 4.

It is necessary to emphasize that Chow (1987d, 1992a, 1992b, 1994, 1995) agrees with the relativistic and SPOPE arguments if they are applied to non-experimental research (or to everyday observations) in which there is no proper control in the technical senses discussed in Section 4.5.1. However, it is Chow's contention that the control provisions in experimentation serve to counteract effectively the relativistic or SPOPE factors. In other words, empirical researchers are capable of adopting (and many do adopt) a detached, impersonal attitude towards knowledge when they conduct experiments.

7.7.2.   The Researcher's Pro Forma Prior Degree of Belief

With the exception of the Bayesian sequential sampling problem, it is necessary to consider the phenomenon-hypothesis consistency (see Section 3.2.1) or the prior data-hypothesis consistency (see Sections 3.2.4 and 7.4.1 above). Seen in this light, the difference between R's and S's degree of belief in Hypothesis H in [P7-7] above may be given the following non-personalist interpretation:

> **The prior data-Hypothesis H consistency is more acceptable to Researcher R than to Researcher S.**
>
> [P7-8]

Proposition [P7-8] may be used to question the putative impact of the prior degree of belief in the hypothesis on empirical research. Recall from Table 5.6 that 'prior data' refers to the to-be-explained Phenomenon P, or to Phenomenon P plus converging research data available to date. Hence, the situation described in Proposition [P7-8] may arise when either (a) Researchers R and S understand Phenomenon P differently, or (b) R and S are both aware of different amounts or kinds of research data to date.

Be that as it may, as objective experimenters, both R and S should give the to-be-tested hypothesis the benefit of the doubt when they test Hypothesis H. That is, for the purpose of testing Hypothesis H, Researchers R or S has to (a) make explicit what their understanding of the hypothesis is, (b) assume that Hypothesis H is true, (c) derive a valid implication from the hypothesis with reference to the experimental task, (d) design and conduct the experiment to test Hypothesis H, and (e) draw the experimental conclusion with the help of inductive and deductive logic.

It is actually difficult to see how Researcher R or S can do otherwise. This is because the derivation of the to-be-tested implication from Hypothesis H is governed by the phenomenon-hypothesis consistency and deductive logic. Nor should the prior degree of belief in the hypothesis affect the conclusion drawn, because the reasoning is constrained by (a) the inductive principle underlying the design of the experiment (see Section 4.5), and (b) the three embedding conditional syllogisms described in Section 4.3. The truth of the matter is that deductive logic and inductive logic do not behave differently for different researchers, regardless of their degrees of belief in the to-be-corroborated hypothesis.

In sum, the to-be-tested hypothesis is always assumed to be true when the experiment is being designed and conducted. This treatment of the hypothesis is represented by using the hypothesis as the antecedent of a conditional proposition (e.g., the antecedent of Proposition [P4.1.2] in Table 4.1). In Bayesian parlance, the *pro forma* prior degree of belief in the hypothesis is always one when it is being tested. The researcher is able to overcome the influences due to the SPOPE factors by virtue of the formal procedures discussed in Chapters 2, 3 and 4.

7.8. The Bayesian Appeal to Confirmation Revisited

To Bayesians, research data are used to confirm hypotheses (in the sense of increasing the posterior degree of belief), and the central question is how non-statistical hypotheses and theories may be confirmed with the Bayesian approach. Given the prominence of the posterior degree of belief, it seems reasonable to expect a Bayesian explication of how the posterior degree of belief contributes to confirmation. Does it mean that the hypothesis which enjoys a higher posterior degree of belief is confirmed to a greater extent than one in which the degree of belief is lower? Although there is no explicit answer to the question, it seems to be an affirmative one, as may be extrapolated from the statement, 'the confirmation . . . hypothesis receives from a positive instance can be represented in terms of an incremental increase in probability . . .' (Earman, 1992, p. 233). Be that as it may, it is problematic to appeal to the confirmation procedure as a means of choosing among contending theories (Popper, 1968a/1959, 1968b/1962). The example Chow (1987d) uses to question the ecological validity argument, as depicted in Table 7.4, may be used to show why the appeal to confirmation is a problematic way to choose between two contending hypotheses.

The to-be-explained phenomenon is that the highways are wet in the morning. Two possible explanations are (1) it rained overnight and (2) someone washed the highways.[5] Suppose that the 'Someone washed the highways' hypothesis is to be tested by accumulating confirming evidence in

Table 7.4   *Two hypotheses of a phenomenon and some of their respective implications*

Phenomenon		Wet highways
Hypotheses	It rained	Someone washed the highways
Potential evidence guided by the confirmation principle	Wet side streets in the city Wet playgrounds in the city Wet uncovered parking lots in the city	
Unambiguous evidence	Wet leaves on top of tall trees	Dry leaves on top of tall trees

deference to Bayesianism. It is inadvisable to collect evidence in situations different from the one in which the to-be-explained phenomenon occurs if the confirmation approach is adopted. Nor should the data be different from the to-be-explained phenomenon. That is, one should not collect ecologically invalid data in Neisser's (1976, 1988) sense of the term.

If ecologically invalid data are used in the confirmation attempt, negative evidence can easily be explained away by pointing out that the original phenomenon and the evidence are different. For this reason, evidence is sought by restricting the observations to objects at the ground level in the example depicted in Table 7.4. Hence, the potential evidence from the confirmation perspective comes from side streets, uncovered playgrounds and parking lots if the confirmation procedure is followed. At the same time, these items of potential evidence are ecologically valid in the sense that they are similar to the original phenomenon. The 'Someone washed the highway' hypothesis is thereby confirmed by each of the wet side streets, wet playgrounds, and wet uncovered parking lots listed in Table 7.4.

However, these observations are not good theory-corroboration evidence for the simple reason that each one of them also confirms a competing hypothesis, 'It rained'. This state of affairs shows that the confirmation strategy is not satisfactory. As none of the observations excludes the 'It rained' hypothesis, it is possible to question Earman's (1992) faith in using data to confirm theories with an eliminatory process. This difficulty with appealing to confirmation also applies to Phillips's (1973) explication of the Bayesian procedure.

Suppose that the Popperian (1968a/1959, 1968b/1962) falsification principle is used to corroborate hypotheses. As has been shown in Section 3.2.2, the first step is to deduce from the explanatory hypothesis what else must occur other than the original phenomenon if the hypothesis is true. The more different this 'what else' is from the original phenomenon, the less ambiguous the evidence is. For this reason, the choice between the two hypotheses in Table 7.4 is less ambiguous if it is based on whether or not the leaves on top of tall trees are wet.

It is true that there is the complication of auxiliary assumptions. Specifically, for the 'Wet leaves' example in Table 7.4 to be useful as evidence for choosing between 'It rained' and 'Someone washed the highways', it is assumed that the atmospheric condition is calm. A windy condition would render dry leaves on top of tall trees ambiguous. However, this complication means that another example is needed because one auxiliary assumption is not justified (Chow, 1989, 1990). This complication does not speak ill of the logic itself because the choice of the new evidence would follow the same principle.

To appeal to leaves on top of tall trees is not replicating the original, or something similar to the original, phenomenon. That is to say, leaves on tree tops in the present example are not ecologically valid data in Neisser's (1976, 1988) sense. Researchers are discouraged from using ecologically invalid observations if the confirmation strategy is used. Yet, appealing to whether or not leaves on top of tall trees are wet renders the evidence less ambiguous in choosing between the two hypotheses.

The reduction in ambiguity is not the result of the fact that wet leaves confirm the 'It rained' hypothesis. Rather, the reduction in ambiguity is brought about by the incompatibility of wet leaves with the 'Someone washed the highway' hypothesis. This exclusion function is achieved by *modus tollens*, not by the Bayesian data-accumulation process. At the same time, this less ambiguous evidence is not ecologically valid evidence. It is for this reason that ecological validity is not a valid criterion for assessing the validity of theory-corroboration data (Chow, 1987d; Mook, 1983).

It may be summarized that the main difficulties with the Bayesian reliance of confirmation as the means of choosing between competing theories are that (a) data which satisfy the confirmation requirement cannot be used to exclude alternative interpretations, and (b) failures to confirm can readily be explained away by the fact that the original phenomenon has not been properly replicated. At the same time, it may be seen that theory-corroboration is achieved by refuting untenable substantive hypotheses via rejecting experimental hypotheses that are inconsistent with research data.

7.9. Summary and Conclusions

Any synopsis of Bayesianism would include the following Bayesian tenets: (1) probability is defined as the degree of belief in the hypothesis; (2) the Bayesian theorem represents the changing of opinion in quantitative form; (3) the crucial question in empirical research is the posterior degree of belief in the hypothesis of interest, p(Hypothesis|Data); (4) the process of inductive inference is represented, as well as justified, by a probability calculus; and (5) the objective of conducting empirical research is to adjust the degrees of belief in hypotheses, not to make the decision of accepting or rejecting hypotheses.

A critique of Bayesianism is of importance in a defence of NHSTP because the mathematical basis of NHSTP and the rationale of empirical research (particularly the hypothetico-deductive character of theory-corroboration research) are questioned by Bayesianism. The first difficulty of Bayesianism is that the issue of phenomenon-hypothesis consistency is neglected. Perhaps the Bayesian view of the nature of empirical research is influenced by the prototype of the empirical question envisaged in Bayesianism, namely, the sequential sampling problem. However, the sequential sampling problem is not a typical empirical research situation.

Bayesians cannot avoid using the frequentist meaning of probability. In fact, they often conflate the frequentist and personalist meanings of probability. In all, Bayesianism is not a convincing alternative approach to empirical research outside the confines of the sequential sampling problem. There is no reason for empirical researchers in general, and experimenters in particular, to give up NHSTP in favour of Bayesian statistics when the experimental situation is not the sequential sampling situation. In the final analysis, the choice between the associated probability and inverse probability seems to be based on the more fundamental question: Are probability statements in empirical research assertions about (a) research data, (b) the tenability of the substantive hypothesis or (c) the subjective degree of belief in the substantive hypothesis? A case has been made that probability statements are statements about research data.

Appendix

That Bayes (1958/1763) recognized the *frequentist* probability may be seen if the problem which first prompted Bayes (1958/1763) to explore the nature of probability is described. This discussion also explains why the Bayesian prototype of empirical research is characterized as 'reflexive'. Bayes (1958/1763) was concerned with the following problem:

> Given the number of times in which an unknown event has happened and failed: Required the **chance** that the **probability** of its happening in a single trial lies somewhere between any two degrees of **probability** that can be named. (Bayes, 1958/1763, p. 298; emphasis in boldface added)
>
>   [A7-1]

At the same time, Bayes defined probability as

> By chance I mean the same as probability. (Bayes, 1958/1763, p. 299)
>
>   [A7-2]

Important to note that the term *probability* has different meanings in [A7-1] and [A7-2]. To properly explicate the two uses of probability, Proposition [A7-1] may be expressed as follows:

(1)   Event E (e.g., the coin lands on its head) is found x times out of N occasions in which E could have occurred (viz., occasions of flipping of the coin).

(2) What is the chance that the 'probability' of Event E occurring on any other occasion on which E can occur (e.g., Occasion N + 1 for ease of exposition)?

(3) The said 'probability' is to be a value within an interval (viz., between v_1 and v_2).

(4) What is the chance of the said 'probability' being between v_1 and v_2?

The question in (4) is the converse of the confidence-interval estimate in traditional statistics. Note that a statement about the confidence-interval estimate is one about a sample statistic. This suggests that, in (2)–(4), Bayes (1958/1763) is dealing with something at the level of abstraction of the sample statistic, not of raw data. Hence, 'probability' in [A7-1] is used to refer to a sample statistic. The statistic in question is a proportion, which is numerically equivalent to a probability. Consequently, the meaning seems clearer if 'probability' in [A7-1] is replaced by 'probability-statistic' or 'proportion'.

The second consideration is that the N occasions in Step (1) are considered as a collective or aggregate. Moreover, what is important about this collective is the proportion, x/N. In stipulating that the statement about 'chance' on Occasion N + 1 be based on what has occurred on the preceding N occasions, Bayes seems to be assuming that Occasion N + 1 is part and parcel of a running collective. In other words, the chance statement is one about the proportion, $x'/(N + 1)$. This proportion is a ratio between two frequencies (viz., x' and N + 1). That is, it is frequentist 'probability' par excellence.

Third, the event in question, E, is not a phenomenon independent of the data-collection procedure. Hence, the 'chance' statement is not about a phenomenon with an independent existence. Instead, it is about the outcome of the ongoing process of data collection. That is, there is no coin landing on its head without tossing the coin. The situation is hence characterized as *reflexive*. Moreover, the outcome is determined by some chance factors found in the data-collection situation. In view of [A7-2], the 'chance' in [A7-1] may be clarified as 'probability-as-chance-influence'.

In short, the proper explication of chance and probability in [A7-1] are 'probability-as-chance-influence' and 'proportion', respectively. Bayes's (1958/1763) problem is to determine the 'probability-as-chance-influence' that the proportion of Event E occurring on the ($N + 1$)th occasion (viz., $x'/(N + 1)$) be between Values v_1 and v_2. Phillips (1973) characterizes Bayes's (1958/1763) task as the 'sequential sampling procedure' (p. 66). However, the characterization does not capture the reflexive feature in Bayes's problem situation.

To iterate, in Bayes's situation, there is no pre-existing phenomenon whose occurrence is independent of the data-collection procedure. Instead, the hypothesis of interest is one about the consequence of the data-collection procedure. In other words, the Bayesian problem is a reflexive one in the sense that it is created by the data-collection procedure itself.

Notes

1. These values represent Editor E's respective degrees of belief in the three hypotheses. This state of affairs amounts to 'mapping the topography of the space of possible theories that cover the explanatory domain in question' (Earman, 1992, p. 183). To Bayesians, these three probabilities need not add up to one. This is not unreasonable because the three hypotheses need not be mutually exclusive. For example, some evidence may be compatible with more than one hypothesis. None the less, the introduction of a new hypothesis (e.g., H_{CL}, the hypothesis that the centre and left parties will form the new coalition government) would bring about a redistribution of probabilities among the new set of four hypotheses (Earman, 1992, p. 229; Phillips, 1973, p. 85).

2. This is, in a sense, an impossible (if not meaningless) task in view of the reflexive features in Bayesianism. As Bayesians do not concern themselves with hypotheses about events independent of their data-collection exercises (recall the example of Editor E), Bayesians do not talk about the tenability of a substantive hypothesis (see Oakes's, 1986, comment quoted in Section 7.5). However, Bayesians do not explicitly restrict Bayesianism to their sequential sampling problem. Instead, Bayesianism is presented as though it is applicable to all empirical research. For this reason, the following argument is presented as though it makes sense to talk about the tenability of the substantive hypothesis in Bayesian parlance.

3. About these objectives, Nagel (1940) advised readers of Jeffreys's (1961/1936) book that 'it is worth noting that in general [Jeffreys] is an unsafe guide in all matters dealing with foundation questions in logic and mathematics' (p. 525).

4. In a slightly different vein, Shafer (1993) talks about the 'ideal picture of probability' (p. 167).

5. For the sake of simplicity, assume that these two alternatives are mutually exclusive for the purpose of this illustration.

8

A Case for NHSTP

NHSTP is faulted by its critics for failing to serve functions it is not developed to do. This unfortunate fate of NHSTP is revealed by the fact that critics envision two prototypes of empirical research, neither of which is suited for theory corroboration. There are reservations about the putative usefulness of the effect size and the function of statistical power. Moreover, the possibility of using the effect size or power analysis depends on (i) changing the meaning of 'Type II error', and (ii) misrepresenting NHSTP at the graphical level. The putative shortcomings of NHSTP are more apparent than real if (a) technical terms are used more carefully, (b) the unwarranted connotations of some terms are avoided, (c) some important distinctions are made, (d) the null hypothesis is used to talk about the data-collection situation, and (e) the rationale of theory corroboration is made explicit. The discussion ends with a recapitulation of some basic issues.

8.1. Introduction

Critics find NHSTP of limited use at best, detrimental to useful empirical research at worst, when they use statistical analyses to determine, first, the practical importance or the real-life impact of the empirical research, and, second, the probability that the to-be-tested hypothesis is true. Consequently, some critics propose (a) to use the confidence-interval estimate or effect size, (b) to determine the sample size with reference to statistical power, and (c) to conduct meta-analysis. Bayesian critics, on the other hand, recommend computing the inverse probability in order to ascertain the probability of the substantive hypothesis.

Contrary to the criticisms levelled against it, NHSTP provides a well-defined rational basis for deciding whether or not research data are due to chance influences. It is true that NHSTP is not informative as to which specific non-chance factor is responsible for the data in the event that chance influences are ruled out. However, the isolation of the non-chance factor is not a statistical concern, but an issue of induction. That is, NHSTP is not meant to be used to interpret data at any level of discourse other than the statistical level. Important to the present discussion is the fact that the inductive method implicated is *not* the 'enumeration plus generation' method. Furthermore, NHSTP and the more sophisticated inductive rules underlying experimental designs are independent components of empirical research. They serve different functions.

It is argued in this defence that the putative ills of using NHSTP cannot be rectified by using the effect size, choosing the sample size with reference to power analysis, conducting meta-analysis or computing the inverse probability. This is the case because it is possible to talk about the effect size or the power of the test only when the meaning of 'Type II error' is changed and NHSTP is graphically misrepresented. The assumptions of the meta-analytic approach to empirical research are debatable. Moreover, the Bayesian posterior degree of belief is meaningful only in the sequential sampling situation that is reflexive in nature. In the final analysis, it may be seen that various criticisms of NHSTP are based on critics' diverse assumptions about the prototype, goal and nature of empirical research. These diverse assumptions are made explicit by identifying the two empirical research prototypes assumed by critics of NHSTP, namely, the utilitarian and the reflexive prediction models.

That NHSTP is misrepresented in the approaches driven by the utilitarian and the sequential sampling prototypes of empirical research identified may be seen from the fact that neither of the two prototypes captures the essence of the theory-corroboration research. NHSTP may be better appreciated by (a) tightening the terminology, (b) avoiding the unwarranted connotative meanings of some technical terms, (c) making some meta-theoretical distinctions, (d) recalling the probabilistic basis of statistical decisions, and (e) considering some issues pertaining to research methodology. The present defence of NHSTP begins with Questions [Q8-1]–[Q8-3]:

What is the role of NHSTP in empirical research?   [Q8-1]

What do critics have in mind when they think of empirical research?   [Q8-2]

Why do critics think empirical research is conducted?   [Q8-3]

It becomes obvious that some of the issues and distinctions may better be seen if they are introduced in question form. Consequently, many further questions are introduced in this summary chapter. Readers may find Table 8.1 helpful, in which all questions are listed in their order of appearance.

8.2.   Two Implicit Prototypes of Empirical Research

The two prototypes of empirical research underlying the criticisms of NHSTP are the utilitarian model and the sequential sampling model. With a few exceptions, absent in criticisms of NHSTP is the consideration of the methodological and conceptual issues implicated in empirical research. A notable example is the neglect of the rationale and requirement of corroborating explanatory hypotheses, in which unobservable theoretical structures are postulated. It is taken for granted in criticisms of NHSTP that the

Table 8.1 *The set of questions related to criticisms of NHSTP arranged in their order of appearance*

Question	
What is the role of NHSTP in empirical research?	[Q8-1]
What do critics of NHSTP have in mind when they think of empirical research?	[Q8-2]
Why do critics of NHSTP think empirical research is conducted?	[Q8-3]
Is there a difference in yields between the two plots or conditions?	[Q8-4]
Is the difference between $\overline{X}_{new}$ and $\overline{X}_{old}$ extreme enough not to be ignored?	[Q8-5]
Is the new fertilizer, or the research treatment, effective (as an efficient cause)?	[Q8-6]
Is the research manipulation effective (as the consequence of a material or formal cause)?	[Q8-6']
What is the magnitude of the effect?	[Q8-7]
How important is the effect?	[Q8-8]
What will be the yields if x amount of the new fertilizer is used?	[Q8-9]
On the basis of what is known about the statistic to date, what is the probability that the statistic has a value between v_1 and v_2?	[Q8-10]
What is the probability that the hypothesis is true, given the data?	[Q8-11]
How do people change their prior opinion when additional data are available?	[Q8-12]
Is the $(\overline{X}_{new} - \overline{X}_{old})$ difference due to chance?	[Q8-13]
Does the answer to [Q8-13] warrant accepting the experimental hypothesis?	[Q8-14]

statistical alternative hypothesis (and, in rare occasions, the null hypothesis itself) is the substantive hypothesis. It is also assumed in the criticisms that to test the statistical hypothesis is to corroborate the substantive hypothesis. Furthermore, no distinction is made between (a) assessing whether or not the research data warrant accepting the substantive hypothesis and (b) evaluating the real-life implications of the research result.

8.2.1. The Utilitarian Model

NHSTP was first introduced in the following context. A new fertilizer was administered to one plot of land, whereas the old fertilizer was applied to another plot of land. This situation suggests Question [Q8-4], which is often asked in the form of Question [Q8-5]:

Is there a difference in yields between the two plots or conditions? [Q8-4]

Is the difference between $\overline{X}_{new}$ and $\overline{X}_{old}$ extreme enough not to be ignored? [Q8-5]

It is important to note that Questions [Q8-4] and [Q8-5] are binary questions (e.g., *a difference* versus *no difference* and *not large enough* versus *large enough*, respectively). Note that either level of the research manipulation (i.e., the new or old fertilizer of *Type of Soil Supplement*) is an efficient cause. Hence, it is easy (as well as meaningful) to be concerned with the magnitude of the difference and the practical implications of the difference so identified. This is the case because this genre of research is conducted for a practical reason, namely, to determine whether or not to use the new fertilizer. Consequently, it is understandable (albeit not justifiable at the conceptual level) that Question [Q8-5] leads to Questions [8-6]–[Q8-9]:

Is the new fertilizer, or the research treatment of interest, effective (as an efficient cause)?   [Q8-6]

What is the magnitude of the effect?   [Q8-7]

How important is the effect?   [Q8-8]

What will be the yields if *x* amount of the new fertilizer is used?   [Q8-9]

Question [Q8-7]–[Q8-9] are non-binary questions. Noting that NHSTP gives only a binary result, critics find NHSTP wanting because they cannot answer any of the non-binary questions. However, this dissatisfaction with NHSTP is unwarranted, as may be seen by considering the relationship between Questions [Q8-5] and [Q8-6].

Question [Q8-5] is a question about data, and it is often phrased as Question [Q8-6′], a form indistinguishable from Question [Q8-6] if the explication in parentheses in both Questions [Q8-6] and [Q8-6′] is not included.

Is the research manipulation effective (as the consequence of a material or formal cause)?   [Q8-6′]

The explications show that there is a subtle difference between Questions [Q8-6] and [Q8-6′]. On the one hand, Question [Q8-6] is asked when the purpose of conducting the research is specifically about the research treat-ment itself. That is, the to-be-investigated phenomenon is reflexive of the data-collection procedure itself. Moreover, the specific concern is the extra-research consequences of the research results. In other words, the emphasis is on the connotation of the word 'effect' in Question [Q8-6], which suggests 'being substantively efficacious'. This use of 'effect' may be called the 'substantive efficacy', and it is the consequence of an efficient cause.

Question [Q8-6′], on the other hand, is raised when the to-be-investigated phenomenon is something independent of the data-collection procedure. That is, neither the research treatment nor real-life consequence is the concern of the research. At the level of statistics, Question [Q8-6′] is a question about data. Hence, the effect implicated in [Q8-6′] refers to an 'extreme enough' difference. This may be called the 'technical effect', and it is the consequence of a material or formal cause.

In other words, there is a conceptual gap going from Question [Q8-4] or [Q8-5] (i.e., the technical effect) to the specific question as to whether or not the substantive manipulation is efficacious (viz., the substantive efficacy in Question [Q8-6]). The conceptual gap becomes more obvious if Question [Q8-5] is considered in the context of Question [Q8-8]. This is the case because it can be seen readily that Questions [Q8-5] and [Q8-8] belong to different domains. It may be necessary to ask [Q8-7] before asking [Q8-8].

The realization of the conceptual gap (between the technical effect and the substantive efficacy) makes explicit two levels of abstraction, namely, (i) the level of data (viz., $X_{new} - X_{old}$), and (ii) the level of the real-life impact of data. However, there are two reasons why this distinction is easily overlooked in utilitarian research. First, the research treatment of interest in the utilitarian prototype is the to-be-studied efficient cause (the new ferti-lizer, in the present example) whose extra-research importance is the reason for conducting the research. Second, 'being efficacious' is a commonly accepted connotative meaning of the term 'effect' in everyday speech. It is a more common practice to think of the consequence of an efficient cause in everyday discourse than of a formal or material cause. The conceptual gap between [Q8-5] and [Q8-6] also shows that, at the conceptual level, a distinction need be made between the to-be-studied phenomenon (e.g., the substantive efficacy of the new fertilizer) and the result of the research about the phenomenon (viz., $\overline{X}_{new}$ and $\overline{X}_{old}$). However, the distinction is obscured in utilitarian research by the reflexivity inherent in the paradigm (i.e., the to-be-investigated phenomenon is the result of the data-collection procedure).

8.2.2. The Sequential Sampling Model

Question [Q8-9] is a prediction problem, a question also found, in a different form, in the Bayesian sequential sampling model. This may be seen from Question [Q8-10], which underlies the Bayesian approach.

On the basis of what is known about the statistic to date, what is the probability that the statistic has a value between ν_1 and ν_2? [Q8-10]

It is important to note that Bayes (1958/1763) had in mind the situation in which there was no to-be-studied phenomenon independent of the data-collection procedure. In the example described in Section 7.2.1, Editor E undertakes a course of action before the election. That is, although Editor E's hypotheses are speculations about the election, the reason of Editor E's conducting the poll is not determined by the election result (viz., whether H_L, H_C or H_R is true). The poll is conducted to justify a course of action that is undertaken before the election.

In other words, Editor E's actual concern is the result of the data-collection operation itself, not the phenomenon about which the hypotheses are postulated. (Hence, the reflexive characterization is also given to the Bayesian sequential sampling procedure described in Section 7.2.3.) The answer to Question [Q8-10] is revised as additional data are accumulated. The Bayesian theorem is an attempt to answer Question [Q8-10].

The Bayesian approach is attractive at first glance for two reasons. First, that people do often change their opinion after receiving new information is indisputable. Bayesianism is presented as a means to quantify this process of changing opinion. Second, users of the Bayesian theorem are promised the probability of the tenability of the substantive hypothesis. Specifically, the result of applying the Bayesian theorem is the posterior probability of the substantive hypothesis. This information is the putative answer to Question [Q8-11]:

What is the probability that the hypothesis is true, given the data? [Q8-11]

NHSTP is found wanting because it cannot answer this question. Bayesians, consequently, fault users of NHSTP for being unable to account for Question [Q8-12]:

How do people change their prior opinion when additional data are available?   [Q8-12]

Question [Q8-12] is an interesting and important question. However, it is not clear how relevant it is as a methodological question. Bayesians have not successfully demonstrated how an individual's certitude about a hypothesis is informative as to the tenability of the hypothesis. Given the fact that Bayesians define 'probability' as the degree of belief in the hypothesis, it is also questionable whether or not Bayesians can really answer Question [Q8-11].

The Bayesian 'change in opinion' refers to a redistribution of degrees of belief among a set of hypotheses. Moreover, the Bayesian hypotheses are hypotheses of different numeral values of the same parameter (e.g., the predicted proportion is x, y or z). This is reasonable in Baye's (1958/1763) situation because this is the only thing that can change the sequential sampling situation. However, in cases where hypotheses are postulated to explain phenomena, the change in opinion is a qualitative matter in the sense that what gets changed is what is said about the phenomenon. This type of change is not numerical in nature.

In sum, Bayesians might be correct to believe in Baye's theorem if the sequential sampling problem epitomizes all empirical research. However, this is not the case. Furthermore, the Bayesian view of empirical research is also debatable for two additional reasons. First, Bayesians are indifferent to the question as to whether or not an explanatory hypothesis is consistent with the phenomenon for which the hypothesis is postulated. Second, in treating the Bayesian theorem as an inductive rule, Bayesians have unjustifiably neglected the difference between empirical research and formalism.

8.3.   Research Objective and Implicit Prototype

It is instructive to examine Questions [Q8-6]–[Q8-9] more closely because they are informative as to the objective of conducting the utilitarian research. The positive answer to Question [Q8-6] is that the research treatment is substantively efficacious. Strictly speaking, [Q8-6] is a meaningful question only if an unambiguously positive answer is given to Question [Q8-5]. Whether or not the positive answer to [Q8-5] is unambiguous is a matter beyond statistics. This may be seen as follows.

Suppose that the answer to Question [Q8-5] is a positive one. By itself, it does not necessarily mean that the difference is really the result of applying the new and old fertilizers to the two plots of land in question. That is, the research manipulation may not be the reason for the difference obtained. For example, the researcher may have failed to take into account the soil, moisture or wind condition. To render Question [Q8-6] meaningful, it is necessary for the researcher to exclude all alternative explanations, except the new fertilizer. Seen in this light, [Q8-6] is not a question about data analysis. Instead, it is one about the design of the data-collection procedure.

In seeking the answer to Question [Q8-7], the researcher assumes that the extra-research consequence of the research manipulation is related, preferably linearly, to the magnitude of $d = (u_E - u_C)/\sigma_E$. The same is true of the answer to Question [Q8-9], which may be sought if it is important to calibrate the extent of the research treatment with reference to its substantive efficacy. The answer to [Q8-8] concerns the evaluation of the research result with reference to a criterion outside the domain of research methodology altogether. In sum, Question [Q8-6] is actually a question about research design. The concerns underlying Questions [Q8-7]–[Q8-9] are concerns with neither research data nor the validity of the research *per se*, but with the practical or utilitarian consequences of research data. In short, Questions [Q8-6]–[Q8-9] have nothing to do with statistics.

It must be emphasized that this is not a denial of the importance of Questions [Q8-6]–[Q8-9]. The issues are rather that (a) Question [Q8-6] is not synonymous with [Q8-5], and (b) NHSTP is not developed to answer Questions [Q8-6]–[Q8-9]. Instead, the outcome of NHSTP is the necessary condition for answering Questions [Q8-6]–[Q8-9].

Given the nature of the sequential sampling problem prototype, the statistical question for Bayesians is the reflexive question about the data-collection procedure itself (viz., Question [Q8-10]). The reflexive question, in turn, is the substantive question. In the case of the utilitarian prototype, the distinctions among the substantive, research and statistical hypotheses are obscured by the fact that the research treatment in utilitarian research is the efficient cause of interest. The research question is also reflexive because the to-be-investigated phenomenon is the result of data-collection procedure itself. Moreover, the research objective is to evaluate the practical importance of the said efficient cause in terms of non-statistical, non-methodological or non-conceptual criteria. Consequently, Bayesian and utilitarian critics of NHSTP have not found it necessary to separate the statistical concern (viz., whether or not the data are the result of chance influences) from pragmatic considerations (viz., those informed by the answers to Questions [Q8-6]–[Q8-10]). However, it seems reasonable to suggest that the onus is on the critics to show why such a separation is not honoured in view of the conceptual gap between Questions [Q8-5] and [Q8-6].

8.4.   Statistical versus Non-statistical Issues

It may be seen from the aforementioned explication of Question [Q8-6]–[Q8-10] that Questions [Q8-13] and [Q8-14] have been neglected in criticisms of NHSTP:

Is the $(\overline{X}_{new} - \overline{X}_{old})$ difference due to chance?   [Q8-13]

Does the answer to Question [Q8-13] warrant accepting the experimental hypothesis?   [Q8-14]

Question [Q8-13] is, in a more general sense, the more informative way to ask Question [Q8-4] or [Q8-5] at the conceptual level. This is the question which makes statistics relevant to empirical research. It is true that Bayesians use the Bayesian theorem to answer Question [Q8-14]. Bayesians treat [Q8-14] as a question about the researcher's certitude. However, as has been argued, knowing the researcher's certitude vis-à-vis the substantive hypothesis is not knowing whether or not the data warrant the certitude, let alone *why* the data warrant the certitude. Bayesians have not considered the validity of the researcher's criterion of certitude. At the same time, it has to be noted that [Q8-14] implicates both inductive and deductive logic. Moreover, the underlying inductive principle is not the 'enumeration plus generalization' induction implicitly assumed in criticisms of NHSTP.

Some points of the present defence of NHSTP are that (a) NHSTP is defensible as the tool to use to answer question [Q8-13], [Q8-4] or [Q8-5]; (b) Question [Q8-14] is answered with the inductive rule underlying the experimental design and the ensuing series of embedding conditional syllogisms (see Section 4.3); (c) NHSTP provides the minor premiss for the innermost of three conditional syllogisms, (d) Question [Q8-14] belongs to a domain different from that implicated in Questions [Q8-6]–[Q8-10]; and (e) [Q8-14] is a theory-corroboration question, not a statistical question.

Question [Q8-13] is a less ambiguous formulation of Question [Q8-4] or [Q8-5] for the following reason. To ask Question [Q8-13] is to make explicit that Question [Q8-5] is asked because of the possibility of chance influences. This possibility is responsible for the ambiguity or uncertainty of the observed $(\overline{X}_{new} - \overline{X}_{old})$ difference. In other words, [Q8-13] and [Q8-6] are questions about explanations of the data at two different levels of discourse. Specifically, chance influences and the research manipulation as the efficient causes are implicated in the answers to Questions [Q8-13] and [Q8-6], respectively.

A surprising feature in criticisms of NHSTP is the indifference to the question of what statistics is used for, namely, to determine whether or not data may be explained in terms of chance influences found in the data-collection procedure. Instead, NHSTP is given functions which are served by (a) the research design, (b) rules for making pragmatic decisions, and (c) criteria for making evaluative judgements in various domains. The crucial point is that critics unjustifiably allow the *content* of the research to overshadow *one formal* requirement of the research. NHSTP may be found satisfactory if it is recalled that conducting empirical research involves considerations and skills belonging to several independent domains. These domains can, and should, be kept distinct when assessing the empirical research, including the evaluation of the role of statistics in empirical research.

8.5. NHSTP in Theory Corroboration

That critics have not distinguished between different components of empirical research may be the consequence of the research prototypes they have in mind when they talk about statistics in general and NHSTP in particular. However, the utilitarian and Bayesian prototypes have not exhausted all empirical research. Consequently, the present defence of NHSTP necessarily leads to a discussion of a different empirical research prototype, namely, the theory-corroboration research.

The non-statistical nature of criticisms of NHSTP are better seen if (a) the three technical meanings of *control* are made explicit, and (b) the distinctions among non-experimental, experimental and quasi-experimental researchers are made with reference to the three meanings of *control*. To make explicit the actual meaning of Question [Q8-5] or [Q8-13] in the face of the various digressions brought about by Questions [Q8-6]–[Q8-11], the consistency between the to-be-explained phenomenon and the explanatory hypothesis in the question is first emphasized (viz., the phenomenon-hypothesis or prior data-implication consistency). Moreover, a quartet of four hypotheses associated with the to-be-explained phenomenon are made explicit, namely: (a) the substantive, explanatory hypothesis, (b) the research hypothesis, (c) the experimental hypothesis, and (d) the implication of the experimental hypothesis at the statistical level (viz., the statistical alternative hypothesis, H_1).

The motivation of the theory-corroboration research is not the pragmatic considerations found in the utilitarian prototype or the Bayesian sequential sampling problem prototype. Instead, the concern is whether or not there is empirical support for the theoretical statements that are used to explain the phenomenon. Chance influences must be ruled out. For this purpose, Question [Q8-13], [Q8-4] or [Q8-5] is first asked. As this is an all-or-none issue, the binary nature of Question [Q8-13], [Q8-4] or [Q8-5] is appropriate.

Given the non-utilitarian nature of the inquiry, it is less likely to treat the $(\overline{X}_1 - \overline{X}_2)$ question (viz., Question [Q8-5]) as a question about the efficient cause (i.e., Question [Q8-6]). Consequently, NHSTP users are justified in not raising Questions [Q8-6]–[Q8-9] when they assess their data. Questions [Q8-6]–[Q8-9] are dealt with after NHSTP. Moreover, entirely different criteria are used to answer these non-statistical questions.

There are five putative difficulties with the experimental theory-corroboration approach. First, there is the issue as to whether or not empirical research is possible at all. This question arises because the explanatory structures or processes postulated in the substantive hypothesis are not observable in the way a physical object is observable (e.g., a chair). The answer to this question is the adoption of the 'conjectures and refutations' approach.

Second, given any to-be-explained phenomenon, there are multiple explanatory hypotheses. The competing hypotheses implicate different hypothetical mechanisms. How is one to choose among these apparently nebulous explanatory alternatives? This difficulty can be resolved by the judicious application of Popper's (1968a/1959, 1968b/1962) 'conjectures and refutations' approach. The researcher has free rein over what hypothesis to propose to account for the to-be-explained phenomenon so long as the hypothesis is consistent with the phenomenon (i.e., the phenomenon-hypothesis consistency). This is the 'conjecture' component of the approach. This is to be corroborated by subjecting the conjecture to as stringent falsification attempts as possible. The qualitatively different attempts constitute collectively a series of converging operations for assessing the tenability of the to-be-corroborated theory or substantive hypothesis. The exclusion function of the 'refutation' component of the approach is rendered possible by using inductive rules.

Third, social constructionists are correct in pointing out that all observations are theory-dependent. This raises the issue as to how it is possible to arrive at non-circular, objective research conclusions. The issue of theory dependence of observations is more apparent than real because the identity of the data does not depend on the to-be-corroborated hypothesis itself. It has been shown that this issue is justifiably resolved with ecologically invalid data.

Fourth, the phenomena of the social psychology of the psychological experiment (or SPOPE) renders objectivity suspect. However, the SPOPE arguments are really criticisms of non-experimental empirical research, not of experimental research. This is the case because the *control* provisions found in properly designed and executed experiments are there to exclude the factors alluded to in the SPOPE arguments.

Fifth, the rationale of theory corroboration depends on a series of embedding conditional syllogisms. Results consistent with the theoretical prescription are used to affirm the consequent of the innermost of the three conditional syllogisms. Results contrary to the theoretical prescription lead to denying the consequent of the innermost of the three conditional syllogisms (i.e., the *modus tollens* rule is used). However, there is an asymmetry between rejecting the consequent of the conditional proposition (i.e., *modus tollens*) and affirming the consequent of the conditional proposition. Specifically, while it is possible to reject unambiguously the antecedent of the conditional proposition when *modus tollens* is used, it is not possible to accept unambiguously the antecedent of the conditional proposition by affirming the consequent. How is this assymmetry resolved? The issues of asymmetry, unobservable mechanisms and competing explanations are dealt with by using deductive logic and inductive logic judiciously at different stages of the empirical research. However, the asymmetry issue can be given a tentative resolution only. It is for this reason that the conclusion of a theory-corroboration experiment is not the closure of a theory-corroboration project. What has to be emphasized is that this failure to provide a closure is not due to any shortcoming of using NHSTP. This state of affairs is dictated by the indeterminacy of affirming the consequent of a conditional proposition. No other statistical index can be used to effect the closure. The best one can hope for is to reduce the ambiguity by (a) using controls in the experiment, and (b) setting up an appropriate series of converging operations for the research project.

8.6. Characteristics of the Present Defence of NHSTP

A common practice found in meta-theoretical discussions of research methodology is to treat deductive logic and inductive logic as mutually exclusive options. This common practice is discarded in the present defence of NHSTP. Instead, it is shown that deductive logic and inductive logic are used at different stages of the theory-corroboration process for different purposes. First, deductive logic is used (a) to deduce from the to-be-tested hypothesis the research hypothesis, and (b) to deduce the experimental hypothesis from the research hypothesis. The statistical alternative hypothesis (H_1) is the implication of the experimental hypothesis at the statistical level. This distinction between H_1 and the substantive hypothesis is an important departure from the usual practice when NHSTP is being discussed. The outcome of NHSTP is used in a disjunctive syllogism to provide the minor premiss for the innermost one of three conditional syllogisms.

Induction is not enumeration plus generalization. Instead, it is a means to exclude alternative explanations of data. A description of the rationale of the more sophisticated rules of induction provides an explication for the three technical meanings of control. These meanings are the criteria for distinguishing among non-experimental, experimental and quasi-experimental studies. This distinction is important because some critics of NHSTP have acknowledged that their criticisms are directed to data collected in situations in which there is no provision for control.

An exploration into this important, but hitherto neglected, caveat shows that most of the criticisms of NHSTP are actually questions about various non-statistical sources of ambiguity in interpreting data collected with non-experimental empirical methods. It is for this reason that the role of NHSTP in empirical research is illustrated in this defence with the role of NHSTP in the theory-corroboration experiment.

Raised in criticisms of NHSTP are also some issues that go beyond research methodology. First, consider the criticism that NHSTP users do not pay attention to the size of the effect of the research manipulation. Behind this criticism is the assumption that there is a quantitative relationship between the size of the effect and the effect's contribution to the tenability of the hypothesis. (For example, larger effects mean presumably greater evidential support.) However, this is not a correct assumption because the null hypothesis is about chance influences on the data-collection procedure, not about the substantive hypothesis. Moreover, the question raised by the null hypothesis is a binary one, namely, *non-chance* versus *chance*. The more important point is that, in the theory-corroboration experiment, the experimental manipulation is neither the substantive manipulation nor the to-be-studied phenomenon.

There is another reason why it is misleading to say that a larger effect size confers more evidential support for the substantive hypothesis. At the level of statistics, the meaning of 'effect' is technically the difference between two conditions (e.g., $d = (u_E - u_C)/\sigma_E$). The said difference has nothing to do with the substantive hypothesis. Moreover, the said difference is the consequence of a formal or a material cause in the theory-corroboration experiment.

Consider the second issue, namely, the justification of the putative contribution of power analysis to empirical research. As the statistical alternative hypothesis is identified with the substantive hypothesis in power analysis, power analysts interpret a significant result to mean that the substantive hypothesis is true. Hence, the power of the test is said to be the probability of the substantive hypothesis being true.

To see why this power analytic argument is debatable, it is necessary to begin with the following definition of 'Type II error':

A Type II error is the error of not rejecting the false null hypothesis.   [D8-1]

It follows from Definition [D8-1] that the probability of Type II error should be the conditional probability shown in the following equation:

$$p(\text{Type II error}) = p(\text{'Retain } H_0\text{'}|\text{not-}H_0)   \quad [\text{E8-1}]$$

However, Equation [8-1] is not useful because the numerical value of the conditional probability represented in [E8-1] is indeterminate. This is the case because the identity of the non-null hypothesis (viz., not-H_0) is not known.

Power analysts do not define the probability of 'Type II error' in terms of H_0. Instead power analysts define it as the probability of 'Type II error$_p$', as follows:

$$p(\text{Type II error}_p) = p(\text{'Retain } H_0\text{'}|H_1)   \quad [\text{E8-2}]$$

Equation [E8-2] is made determinate because an explicit assumption is made about the size of the effect. This assumption renders it possible to have a pair of normal distributions with a well-defined distance between them. With the problem of indeterminacy resolved, power analysts define 'power' as follows:

$$\text{power} = 1 - p(\text{Type II error}_p) = 1 - p(\text{'Retain } H_0\text{'}|H_1)   \quad [\text{E8-3}]$$

There are three problems with this treatment of the power of the test. First, to use Equation [E8-2] is to change the definition of 'Type II error' to 'Type II error$_p$.' The second difficulty is a more serious one, and it may be explicated in terms of the two-sample t test. The statistical significance is determined with one sampling distribution only – namely, the sampling distribution of differences – not two distributions. The mean of this lone sampling distribution is zero (i.e., the mean difference).

There may be reasons to expect a mean difference larger than zero (e.g., 5). However, this situation simply means that the numerator of the t statistic changes from '$(\overline{X}_1 - \overline{X}_2) - (u_1 - u_2) = 0$' to '$\overline{X}_1 - \overline{X}_2) - (u_1 - u_2) = 5$' or '$(\overline{X}_1 - \overline{X}_2) - (u_1 - u_2) - 5 = 0$'. The resultant sampling distribution is displaced more to the right along the horizontal axis which represents all possible values of the difference between two means. That is, only one sampling distribution of differences is still being used. In short, it is misleading to use two distributions (i.e., one for H_0 and the other for H_1) to represent the rationale of NHSTP.

The third difficulty with the power analysis is a conceptual one. The power of the test would still not be the probability that the substantive hypothesis were true even if (a) the aforementioned problem of graphical misrepresentation could be ignored, and (b) H_1 were indeed the substantive hypothesis. Recall from Equation [E8-3] that 'power' refers to the complement of a conditional probability. Specifically, H_1 is assumed correct *before* the numerical value of Type II error becomes available. How can the power of the test be the probability of the truth of H_1 when it is necessary to assume first that H_1 is true?

8.7. Summary and Conclusions

The spirit of many of the criticisms of NHSTP is about good empirical research, something all NHSTP users endorse. At a minimum an empirical research is good if it has statistical conclusion validity and inductive conclusion validity. Many criticisms of NHSTP are actually questions about inductive conclusion validity, whereas NHSTP is relevant only to the statistical conclusion validity. Some other criticisms of NHSTP belong to domains other than research methodology altogether.

For example, questions about the efficacy of the research manipulation are linked to issues about the practical impact of research data. This is obviously an important question. However, it should be raised *before* or *after* the empirical research. It should not be raised in the course of designing, conducting and assessing whether or not data confer evidential support for the substantive hypothesis. Moreover, in the case of the theory-corroboration research, there is no quantitative relationship between the magnitude of $d = (u_E - u_C)/\sigma_E$ and the amount of evidential support for the to-be-corroborated hypothesis conferred by the data once it is determined that the data are too unlikely to be the result of chance influences.

The motivation of power analysis or the Bayesian posterior probability is understandable. It seems reassuring if the researcher can determine the probability of the hypothesis being true in the face of multiple theoretical alternatives. However, whether or not data from Study S warrant accepting the to-be-corroborated explanatory hypothesis is an all-or-none matter. It is not a matter of probability. At the level of research programme, the tenability of the to-be-corroborated hypothesis is established when recognized alternative explanatory hypotheses are justifiably excluded. This is not a quantitative issue at the statistical level, but a qualitative one at the conceptual level.

The limited, but important, role of NHSTP may be better appreciated if certain distinctions are made. First, empirical research is more than statistics. Second, theoretical corroboration is more than statistical hypothesis testing. Third, there are fundamental differences between experimental and non-experimental empirical research. It is also important not to lose sight of the fact that the evidential contribution of research data is neither a numerical nor a probabilistic matter. Consequently, neither the effect size nor statistical power is informative as to the evidential support that research data confer to the substantive hypothesis.

The evidential support for the substantive hypothesis is to be judged by the extent to which alternative explanations to the to-be-corroborated hypothesis can be excluded. This exclusion function is served by inductive rules more sophisticated than the 'enumeration plus generalization' view of induction. The inductive rule is realized by the design of the empirical research. It is for this reason that experimentation is the more satisfactory theory-corroboration method.

Statistical significance means nothing more than the decision that chance influences may be ruled out as an explanation of the data with reference to a particular criterion of strictness. This criterion owes its objectivity and probabilistic characteristics to the sampling distribution of the test statistic. In other words, the null hypothesis is one about the data-collection procedure. Contrary to a widely held belief, the null hypothesis can (and should) be true if (a) the design of the empirical research is correct, and (b) the to-be-corroborated hypothetical structure does not have the property attributed to it.

In the case of the theory-corroboration experiment, whether or not the research data confer any evidential support for the substantive hypothesis is determined by a series of three embedding conditional syllogisms. Statistical significance (or the lack of it) provides the minor premiss necessary for initiating the said series of conditional syllogisms. In short, a case can be made for the validity and usefulness of NHSTP in empirical research, particularly the theory-corroboration experiment. The rationale of NHSTP is sound. At the same time, it must be emphasized that the role of NHSTP in empirical research is a limited one, restricted to deciding whether or not research data can be explained in terms of chance influences. However, it is an important role.

References

Badia, P., Haber, A., & Runyon, R. P. (eds.) (1970). *Research problems in psychology*. Reading, Mass.: Addison-Wesley Publishing Co.

Bakan, D. (1966). The test of significance in psychological research. *Psychological Bulletin*, 66, 423–37.

Bayes, T. (1958). An essay towards solving a problem in the doctrine of chance (originally published in 1763). *Biometrika*, 45, 293–315.

Berkowitz, L., & Donnerstein, E. (1982). External validity is more than skin deep: some answers to criticisms of laboratory experiments. *American Psychologist*, 37, 245–57.

Binder, A. (1963). Further considerations on testing the null hypothesis and the strategy and tactics of investigating theoretical models. *Psychological Review*, 70, 107–15.

Boring, E. G. (1954). The nature and history of experimental control. *American Journal of Psychology*, 67, 573–89.

Boring, E. G. (1969). Perspective: artifact and control. In R. Rosenthal & R. L. Rosnow (eds.), *Artifacts in behavioral research* (pp. 1–11). New York: Academic Press.

Brenner-Golomb, N. (1993). R. A. Fisher's philosophical approach to inductive inference. In G. Keren & C. Lewis (eds.), *A handbook for data analysis in the behavioral sciences*: *Methodological issues* (pp. 283–307). Hillsdale, NJ: Lawrence Erlbaum.

Bridgman, P. W. (1936). *The nature of physical theory*. New York: Dover Publications.

Bridgman, P. W. (1961). *The logic of modern physics* (originally published in 1927). New York: Macmillan.

Campbell, D. T., & Stanley, J. C. (1966). *Experimental and quasi-experimental designs for research*. Chicago: Rand McNally.

Chomsky, N. (1957). *Syntactic structures*. The Hague: Mouton.

Chow, S. L. (1985). Iconic store and partial report. *Memory & Cognition*, 13, 256–64.

Chow, S. L. (1987a). *Experimental psychology*: *rationale, procedures and issues*. Calgary: Detselig.

Chow, S. L. (1987b). Some reflections on Harris and Rosenthal's thirty-one meta-analyses. *Journal of Psychology*, 121, 95–100.

Chow, S. L. (1987c). Meta-analysis of pragmatic and theoretical research: a critique. *Journal of Psychology*, 121, 259–71.

Chow, S. L. (1987d). Science, ecological validity, and experimentation. *Journal for the Theory of Social Behaviour*, 17, 181–94.

Chow, S. L. (1988). Significance test or effect size? *Psychological Bulletin*, 103, 105–10.

Chow, S. L. (1989). Significance tests and deduction: reply to Folger (1989). *Psychological Bulletin*, 106, 161–5.

Chow, S. L. (1990). In defense of Popperian falsification. *Psychological Inquiry*, 1, 147–9.

Chow, S. L. (1991a). Conceptual rigor versus practical impact. *Theory & Psychology*, 1, 337–60.

Chow, S. L. (1991b). Rigor and logic: a response to comments on 'conceptual rigor'. *Theory & Psychology*, 1, 389–400.

Chow, S. L. (1991c). Some reservations about statistical power. *American Psychologist*, 46, 1088–9.

Chow, S. L. (1992a). *Research methods in psychology*: *a primer*. Calgary: Detselig.

Chow, S. L. (1992b). Acceptance of a theory: justification or rhetoric? *Journal for the Theory of Social Behaviour*, 22, 447–74.

Chow, S. L. (1992c). Positivism and cognitive psychology: a second look. In C. W. Tolman (ed.), *Positivism in psychology: historical and contemporary problems* (pp. 119–44). New York: Springer-Verlag.

Chow, S. L. (1994). The experimenter's expectancy effect: a meta-experiment. *Zeitschrift für Pädagogische Psychologie* (German Journal of Educational Psychology), 8, 89–97.

Chow, S. L. (1995). In defense of experimental data in a relativistic milieu. *New Ideas of Psychology*, 13, 259–79.

Cohen, J. (1965). Some statistical issues in psychological research. In B. B. Wolman (ed.), *Handbook of clinical psychology* (pp. 95–121). New York: McGraw-Hill.

Cohen, J. (1987). *Statistical power analysis for the behavioral sciences* (revised edition). New York: Academic Press.

Cohen, J. (1990). Things I have learned (so far). *American Psychologist*, 45, 1304–12.

Cohen, J. (1992a). Statistical power analysis. *Current Directions in Psychological Science*, 1, 98–105.

Cohen, J. (1992b). A power primer. *Psychological Bulletin*, 112, 155–9.

Cohen, J. (1994). The earth is round ($p < .05$). *American Psychologist*, 49, 997–1003.

Cohen, J., & Cohen, P. (1975). *Applied multiple regression/correlation analysis for the behavioral sciences*. New York: John Wiley.

Cohen, M. R., & Nagel, E. (1934). *An introduction to logic and scientific method* London: Routledge & Kegan Paul.

Cook, T. D., & Campbell, D. T. (1979). *Quasi-experimentation: Design and analysis issues for field settings*. Chicago: Rand McNally.

Cook, T. D., & Leviton, L. C. (1980). Reviewing the literature: a comparison of traditional methods with meta-analysis. *Journal of Personality*, 48, 449–72.

Cooper, H. M. (1979). Statistically combining independent studies: a meta-analysis of sex differences in conformity research. *Journal of Personality and Social Psychology*, 37, 131–46.

Cooper, H. M., & Rosenthal, R. (1980). Statistical versus traditional procedures for summarizing research findings. *Psychological Bulletin*, 87, 442–9.

Copi, I. (1965). *Symbolic logic* (2nd edition). New York: Macmillan.

Copi, I. (1982). *Symbolic logic* (6th edition). New York: Macmillan.

Cotton, J. W. (1955). On making predictions from Hull's theory. *Psychological Review*, 62, 303–14.

Cozby, P. C. (1989). *Methods in behavioral research* (4th edition). Palo Alto, CA: Mayfield.

Danziger, K. (1990). *Constructing the subject: Historical origins of psychological research*. Cambridge: Cambridge University Press.

Darlington, R. B., & Carlson, P. M. (1987). *Behavioral statistics: Logic and methods*. New York: Collier Macmillan Publishers.

Earman, J. (1992). *Bayes or bust? A critical examination of Bayesian confirmation theory*. Cambridge, Mass.: MIT Press.

Edwards, W., Lindman, H., & Savage, L. J. (1963). Bayesian statistical inference for psychological research. *Psychological Review*, 70, 193–242.

Eysenck, H. J. (1978). An exercise in mega-silliness. *American Psychologist*, 33, 517.

Falk, R., & Greenbaum, C. W. (1995). Significance tests die hard: the amazing persistence of a probabilistic misconception. *Theory & Psychology*, 5, 75–98.

Ferris, C. D., Grubbs, F. E., & Weaver, C. L. (1946). Operating characteristics for the common statistical tests for significance. *Annual of Mathematical Statisticians*, 17, 178–97.

Fillmore, C. J. (1968). The case for case. In E. Bach & R. T. Harms (eds.), *Universals in linguistic theory* (pp. 1–90). New York: Rinehart & Winston.

Fisher, R. A. (1959). *Statistical methods and scientific inference* (2nd edition). New York: Hafner Publishing Co.

Fisher, R. A. (1960). *The design of experiments* (7th edition). New York: Hafner Publishing Co.

Fodor, J. A. (1975). *Language of thought*. New York: Thomas Y. Crowell Co.

Folger, R. (1989). Significance tests and the duplicity of binary decisions. *Psychological Bulletin*, 106, 155–60.

Frick, R. W. (1995). Accepting the null hypothesis. *Memory & Cognition*, 23, 132–8.

Gallo, P. S., Jr (1978). Meta-analysis: a mixed meta-phor? *American Psychologist*, 33, 515–17.

Garner, W. R., Hake, H. W., & Eriksen, C. W. (1956). Operationalism and the concept of perception. *Psychological Review*, 63, 149–59.

Gergen, K. J. (1991). Emerging challenges for theory and psychology. *Theory & Psychology*, 1, 13–35.

Gigerenzer, G. (1993). The superego, the ego, and the id in statistical reasoning. In G. Keren & C. Lewis (eds.), *A handbook for data analysis in the behavioral sciences: methodological issues* (pp. 311–39). Hillsdale, NJ: Lawrence Erlbaum Associates.

Glass, G. V. (1976). Primary, secondary and meta-analysis of research. *Educational Researcher*, 5, 3–8.

Glass, G. V. (1978). Integrating findings: the meta-analysis of research. *Review of Research in Education*, 5, 351–79.

Glass, G. V., & Kliegl, R. M. (1983). An apology for research integration in the study of psychotherapy. *Journal of Consulting and Clinical Psychology*, 51, 28–41.

Glass, G. V., McGaw, B., & Smith, M. L. (1981). *Meta-analysis in social research*. Beverly Hills, CA: Sage.

Goldstein, J. J., Rosnow, R. L., Goodstadt, B., & Suls, J. M. (1972). The 'good' subject in verbal operant conditioning research. *Journal of Experimental Research in Personality*, 6, 29–33.

Grant, D. A. (1962). Testing the null hypothesis and the strategy and tactics of investigating theoretical models. *Psychological Review*, 69, 54–61.

Gravetter, F. J., & Wallnau, L. B. (1996). *Statistics for the behavioral sciences*. Minneapolis: West Publishing Co.

Green, D. M., & Swets, J. A. (1966). *Signal detection theory and psychophysics*. New York: John Wiley.

Greenwald, A. G. (1993). Consequences of prejudice against the null hypothesis. In G. Keren, & C. Lewis (eds.), *A handbook for data analysis in the behavioral sciences: methodological issues* (pp. 419–48). Hillsdale, NJ: Lawrence Erlbaum.

Harris, M. J. (1991). Significance tests are not enough: the role of effect-size estimation in theory corroboration. *Theory & Psychology*, 1, 375–82.

Harris, M. J., & Rosenthal, R. (1985). Mediation of interpersonal expectancy effects: 31 meta-analyses. *Psychological Bulletin*, 97, 363–86.

Hays, W. L. (1963). *Statistics for psychologists*. New York: Holt, Rinehart & Winston.

Hays, W. L. (1994). *Statistics* (5th edition). Fort Worth, Texas: Harcourt Brace College Publishers.

Hogben, L. (1957). *Statistical theory: the relationship of probability, credibility and error*. New York: W. W. Norton.

Hopkins, K. D., Hopkins, B. R., & Glass, G. V. (1996). *Basic statistics for the behavioral sciences* (3rd edition). Boston: Allyn and Bacon.

Hull, C. L. (1943). *Principles of behavior: an introduction to behavior theory*. New York: Appleton-Century.

Hurlburt, R. T. (1993). *Comprehending behavioral statistics*. Pacific Grove, CA: Brooks/Cole.

Inman, H. F. (1994). Karl Pearson and R. A. Fisher on statistical tests: a 1935 exchange from *Nature*. *American Statistician*, 48, 2–11.

Jeffreys, H. (1961). *Theory of probability* (originally published in 1939). Oxford: Clarendon Press.

Kerlinger, F. N. (1964). *Foundations of behavioral research*. New York: Holt, Rinehart & Winston.

Kirk, R. E. (1984). *Basic statistics* (2nd edition). Pacific Grove, CA: Brooks/Cole.

Koch, S. (1944). Hull's *Principle of Behavior*: a special review. *Psychological Bulletin*, 41, 268–86.

Koch, S. (1954). Clark L. Hull. In W. K. Estes, S. Koch, K. MacCorquodale, P. E. Meehl, C.

G. Mueller, Jr., W. W. Schoenfeld, & W. S. Verplanck (eds.). *Modern learning theory* (pp. 1–101). New York: Appleton-Century-Croft.

Kraemer, H. C., & Thiemann, S. (1987). *How many subjects? Statistical power analysis in research*. Newbury Park, CA: Sage.

Kuhn, T. S. (1970). *The structure of scientific revolutions* (2nd edition, enlarged). Chicago: University of Chicago Press.

Leviton, L. C., & Cook, T. D. (1981). What differentiates meta-analysis from other forms of review? *Journal of Personality*, 49, 231–6.

Loomis, L. R. (ed.) (1971). *Aristotle: on man in the universe*. Roslyn, NY: the Classic Club.

Lykken, D. T. (1968). Statistical significance in psychological research. *Psychological Bulletin*, 70, 151–9.

MacKay, D. G. (1993). The theoretical epistemology: a new perspective on some long-standing methodological issues in psychology. In G. Keren & C. Lewis (eds.), *A handbook for data analysis in the behavioral sciences: methodological issues* (pp. 229–55). Hillsdale, NJ: Lawrence Erlbaum.

Macmillan, N. A. (1993). Signal detection theory as data analysis method and psychological decision model. In G. Keren & C. Lewis (eds.), *A handbook for data analysis in the behavioral sciences: methodological issues* (pp. 21–57). Hillsdale, NJ: Lawrence Erlbaum.

Manicas, P. T., & Secord, P. F. (1983). Implications for psychology of the new philosophy of science. *American Psychologist*, 38, 399–413.

McNicol, D. (1972). *A primer of signal detection theory*. London: George Allen & Unwin.

Meehl, P. E. (1967). Theory-testing in psychology and physics: a methodological paradox. *Philosophy of Science*, 34, 103–15.

Meehl, P. E. (1978). Theoretical risks and tabular asterisks: Sir Karl, Sir Ronald, and the slow progress of soft psychology. *Journal of Consulting and Clinical Psychology*, 46, 429–71.

Meehl, P. E. (1990). Appraising and amending theories: the strategy of Lakatosian defense and two principles that warrant it. *Psychological Inquiry*, 1, 108–41.

Merikle, P. M. (1980). Selection from visual persistence by perceptual groups and category membership. *Journal of Experimental Psychology, General*, 3, 279–95.

Mill, J. S. (1973). *A system of logic: ratiocinative and inductive*. Toronto: University of Toronto Press.

Miller, G. A. (1956). The magical number seven, plus or minus two: some limits on our capacity for processing information. *Psychological Review*, 63, 81–97.

Miller, G. A. (1962). Some psychological studies of grammar. *American Psychologist*, 17, 748–62.

Mintz, J. (1983). Integrating research evidence: a commentary on meta-analysis. *Journal of Consulting and Clinical Psychology*, 51, 71–5.

Mook, D. G. (1983). In defense of external invalidity. *American Psychologist*, 38, 379–87.

Morrison, D. E., & Henkel, R. E. (eds.) (1970). *The significant test controversy: a reader*. Chicago: Aldine.

Mosteller, F., & Bush, R. R. (1954). Selected quantitative techniques. In G. Lindzey (ed.), *Handbook of Social Psychology*. Volume 1: *Theory and Method* (pp. 289–334). Reading, Mass.: Addison-Wesley.

Nagel, E. (1940). Review of 'Theory of Probability'. *Journal of Philosophy*, 37, 524–8.

Neisser, U. (1967). *Cognitive psychology*. New York: Appleton-Century-Croft.

Neisser, U. (1976). *Cognition and reality*. San Francisco: W. H. Freeman.

Neisser, U. (1988). New vistas in the study of memory. In U. Neisser & E. Winograd (eds.), *Remembering reconsidered: ecological approaches to the study of memory* (pp. 1–10). Cambridge: Cambridge University Press.

Neyman, J., & Pearson, E. S. (1928). On the use and interpretation of certain test criteria for purposes of statistical inferences (Part I). *Biometrika*, 20A, 175–240.

Nunnally, J. (1960). The place of statistics in psychology. *Educational and Psychological Measurement*, 20, 641–50.

Oakes, M. (1986). *Statistical inference: a commentary for the social and behavioral sciences*. Chichester: John Wiley & Sons.

Orne, M. T. (1962). On the social psychology of the psychological experiment: with particular reference to demand characteristics and their implications. *American Psychologist*, 17, 776–83.

Orne, M. T. (1969). Demand characteristics and the concept of quasi-controls. In R. Rosenthal & R. L. Rosnow (eds.), *Artifact in Behavioral Research* (pp. 143–79). New York: Academic Press.

Orne, M. T., & Evans, E. J. (1965). Social control in the psychological experiment: antisocial behavior and hypnosis. *Journal of Personality and Social Psychology*, 1, 189–200.

Phillips, L. D. (1973). *Bayesian statistics for social scientists*. London: Nelson.

Pollard, P. (1993). How significant is 'significance'? In G. Keren & C. Lewis (eds.), *A handbook for data analysis in the behavioral sciences: methodological issues* (pp. 448–60). Hillsdale, NJ: Lawrence Erlbaum.

Popper, K. R. (1968a). *The logic of scientific discovery* (originally published in 1959). New York: Harper & Row.

Popper, K. R. (1968b). *Conjectures and refutations* (originally published in 1962). New York: Harper & Row.

Presby, S. (1978). Overly broad categories obscure important differences between therapies. *American Psychologist*, 33, 514–15.

Rachman, S., & Wilson, G. T. (1980). *The effects of psychological therapy*. Oxford: Pergamon Press.

Rosenthal, R. (1963). On the social psychology of the psychological experiment: the experimenter's hypothesis as unintended determinant of experimental results. *American Scientist*, 51, 268–83.

Rosenthal, R. (1973). The Pygmalion effect lives. *Psychology Today, September*, 56–63.

Rosenthal, R. (1976). *Experimenter effects in behavioral research* (enlarged edition). New York: Irvington Publishers.

Rosenthal, R. (1983). Assessing the statistical and social importance of the effects of psychotherapy. *Journal of Consulting and Clinical Psychology*, 51, 4–13.

Rosenthal, R. (1984). *Meta-analytic Procedures for Social Research*. Beverly Hills, CA: Sage.

Rosenthal, R., & Fode, K. L. (1963a). Three experiments in experimenter bias. *Psychological Reports*, 12, 491–511.

Rosenthal, R., & Fode, K. L. (1963b). The effect of experimenter bias on the performance of the albino rat. *Behavioral Science*, 8, 183–9.

Rosenthal, R., & Rosnow, R. L. (1975). *The volunteer subject*. New York: John Wiley.

Rosenthal, R., & Rubin, D. B. (1979). A note on percent variance explained as a measure of the importance of effects. *Journal of Applied Social Psychology*, 9, 395–6.

Rosenthal, R., & Rubin, D. B. (1982). A simple, general purpose display of magnitude of experimental effect. *Journal of Educational Psychology*, 74, 166–9.

Rosnow, R. L., & Rosenthal, R. (1989). Statistical procedures and the justification of knowledge in psychological science. *American Psychologist*, 44, 1276–84.

Rozeboom, W. W. (1960). The fallacy of the null-hypothesis significance-test. *Psychological Bulletin*, 57, 416–28.

Savin, H. B., & Perchonock, E. (1965). Grammatical structure and the immediate recall of English sentences. *Journal of Verbal Learning and Verbal Behavior*, 4, 348–53.

Schmidt, F. L. (1992). What do data really mean? Research findings, meta-analysis, and cumulative knowledge in psychology. *American Psychologist*, 47, 1173–81.

Schmidt, F. L. (1994). Quantitative methods and cumulative knowledge in psychology: implications for the training of researchers. Presidential Address presented to the Division of Statistics, Measurement and Evaluation (Division 5 of APA) at the 102nd Annual Convention of the American Psychological Association, 13 August 1994, Los Angeles.

Schmidt, F. L. (1996). Statistical significance testing and cumulative knowledge in psychology: implications for training researchers. *Psychological Methods*, 1, 115–29.

Schneider, W., & Shiffrin, R. M. (1977). Controlled and automatic human information processing. I: Detection, search, and attention. *Psychological Review*, 84, 1–66.

Serlin, R. C., & Lapsley, D. K. (1985). Rationality in psychological research: the good-enough principle. *American Psychologist*, 40, 73–83.

Serlin, R. C., & Lapsley, D. K. (1993). Rational appraisal of psychological research and the good enough principle. In G. Keren & C. Lewis (eds.), *A handbook for data analysis in the behavioral sciences: methodological issues* (pp. 199–228). Hillsdale, NJ: Lawrence Erlbaum.

Shafer, G. (1993). Can the various meaning of probability be reconciled? In G. Keren & C. Lewis (eds.), *A handbook for data analysis in the behavioral sciences: methodological issue* (pp. 165–96). Hillsdale, NJ: Lawrence Erlbaum.

Siegel, S. (1956). *Non-parametric statistics for the behavioral sciences*. New York: McGraw-Hill.

Skinner, B. F. (1938). *The behavior of organisms: an experimental analysis*. New York: Appleton-Century.

Sohn, D. (1980). Critique of Cooper's meta-analytic assessment of the findings of sex differences in conformity behavior. *Journal of Personality and Social Psychology*, 39, 1215–21.

Sperling, G. (1960). The information available in brief visual presentations. *Psychological Monographs*, 74 (11): entire issue.

Swain, J. F., Rouse, I. L., Curley, C. B., & Sacks, F. M. (1990). Comparison of the effects of oat bran and low-fiber wheat on serum lipoprotein levels and blood pressure. *New England Journal of Medicine, no.* 322, 147–52.

Swanson, J. M., & Kinsbourne, M. (1976). Stimulant-related state-dependent learning in hyperactive children. *Science, no.* 182, 1354–7.

Tukey, J. W. (1960). Conclusions vs. decisions. *Technometrics*, 2, 1–11.

Turner, M. B. (1967). *Psychology and the philosophy of science*. New York: Appleton-Century-Croft.

Wilson, G. T., & Rachman, S. J. (1983). Meta-analysis and the evaluation of psychotherapy outcome: limitations and liabilities. *Journal of Consulting and Clinical Psychology*, 51, 54–64.

Wilson, W. R., & Miller, H. (1964). A note on the inconclusiveness of accepting the null hypothesis. *Psychological Review*, 71, 238–42.

Wilson, W. R., Miller, H., & Lower, J. S. (1967). Much ado about the null hypothesis. *Psychological Bulletin*, 67, 188–97.

Yngve, V. (1960). A model and an hypothesis for language structure. *Proceedings of the American Philosophical Society*, 104, 444–66.

Author Index

Subject Index